KT-464-446

Aberdeenshire
3186981

NIGHTINGALES IN NOVEMBER

NIGHTINGALES IN NOVEMBER

A Year in the Lives of Twelve British Birds

Mike Dilger

B L O O M S B U R Y
LONDON • OXFORD • NEW YORK • NEW DELHI • SYDNEY

Bloomsbury Natural History
An imprint of Bloomsbury Publishing Plc

50 Bedford Square
London
WC1B 3DP
UK

1385 Broadway
New York
NY 10018
USA

www.bloomsbury.com

First published 2016

Maps on pages 45, 107 and 250 redrawn from those appearing on the BTO website, www.bto.org, and derived from data collected by its network of 50,000 volunteers and research staff.

British Library Cataloguing-in-Publication Data
A catalogue record for this book is available from the British Library.

Library of Congress Cataloguing-in-Publication data has been applied for.

ISBN: HB: 978-1-4729-1535-1
ePub: 978-1-4729-1536-8

2 4 6 8 10 9 7 5 3 1

Typeset in Bembo Std by Deanta Global Publishing Services, Chennai, India
Printed and bound in Great Britain by CPI Group (UK) Ltd, Croydon CR0 4YY

To Zachary – for bringing immense joy into our lives

Contents

Introduction 9

The Species 13

January 21

February 49

March 71

April 99

May 129

June 165

July 195

August 223

September 249

October 275

November 305

December 333

Further Reading 360

Acknowledgements 362

Index 364

Introduction

Operating as a small set of islands, located at a temperate latitude and positioned just off the edge of a huge continental land mass, the British Isles are blessed with both clearly marked seasons and a wonderfully varied climate. Despite temperatures varying by 25°C between January and August, and daylight hours more than doubling between the winter and summer solstices, our predictably unpredictable climate never quite approaches the searing heat, mind-numbing cold or seasonal deluges regularly experienced in other corners of the world. The weather, however, must never be taken for granted, and the constantly changing seasons have a profound effect on all who live here. Of course we humans can insulate ourselves from the worst excesses of the weather – by opening our windows in the summer and turning on the lights during the dark days of winter – but our wildlife doesn't have that luxury.

With spring and summer a time for pairing, mating and rearing in Britain, autumn representing the season of preparation, winter then becomes the time for our wildlife to either hibernate, in an effort to bypass the season completely, or hunker down in full survival mode. Unable to hibernate like our amphibians, reptiles and many mammals, the power of flight has given birds another option – that of simply migrating abroad until conditions improve. Leaving behind those resident birds willing to take their chances here all year round, the 'fair weather' summer migrants will simply be replaced as they leave by hardier winter visitors arriving from much colder continental climates. The warming influence of the Gulf Stream and the North Atlantic Drift will ensure those birds spending the winter in Britain are able to positively thrive in the relatively benign conditions offered here, while the long and warm summer months will ensure plenty of time in which to raise broods. When we also factor in the immense array of habitats here – from forests to farmland and estuaries to islands – it is perhaps no surprise that Britain punches well above its weight in terms of biological diversity.

How and when nature responds to the changing of the seasons is in fact a fully-fledged science. Phenology, in plain English, is the study of recurring biological events, such as flowering and leaf fall in plants, and the hibernation and migration of animals. Certainly in Britain, this kind of data has historically been gathered on little more than an ad hoc basis. However, recently it has begun to be collated in a far more systematic manner, now that its roll in understanding how climate change affects the natural world is better understood. Of course wandering around the Staffordshire countryside as a child in the 1970s and 1980s I was certainly unaware of the term 'phenology'. But as someone who has always been happiest outdoors, I intuitively understood that Robins were one of the very few birds to sing on Christmas Day, the appearance of a Cuckoo was always the sign that spring had arrived and hooting Tawny Owls meant the descent into winter had begun.

It's also a measure of the unique popularity of birds in Britain that I am by no means alone in taking great pleasure from noting down my first Swallow in April, or the precise week in May that Blue Tits start bringing caterpillars back to their brood. I'd like to think a substantial number of us effectively adopt 'nature's calendar' to signpost our year rather than perhaps using the human-orientated dates typical throughout a more traditional calendar. The very fact that the Royal Society for the Protection of Birds (RSPB) has over 1.1 million members and the British Trust for Ornithology (BTO) rings around a million birds each year certainly shows that both birdwatching and the amateur study of birds have grown into hugely popular pastimes.

Right now is also an exciting time to be a bird enthusiast, as cutting-edge technology and advances in filming and photography are revealing more about our feathered friends than we ever thought possible. So it is with these facts in mind that I have attempted to generate this alternative almanac. Compiled from monographs, academic papers, my own knowledge and the most recent scientific developments, I have hopefully produced, in an easily digestible form, a slightly different way of marking the year – through the joy of birds.

I believe the twelve birds chosen represent the best of British. I've picked iconic birds that the vast majority of readers will be instantly familiar with and are most likely to see or hear somewhere, at some point during the year. For several of the twelve, the furthest you will have to search is your own back garden or local park, while others may require a bit more effort to catch up with, due to geographical and habitat preferences, or a certain level of rarity. I've selected species with varying and contrasting life histories, so in addition to year-round residents, both summer and winter visitors have been included. Whereas the allegiance of Robins, Tawny Owls and Blue Tits to the Union Jack flag is assured, the nationality of some birds, such as Cuckoos, Puffins and Nightingales, should perhaps now be questioned. Does, for example, being here for little more than six weeks of the year entitle Cuckoos to British citizenship?! Certainly those birds just spending part of their year in Britain won't see international borders or frontiers in quite the same way we do. Instead they might recognise the Sahara Desert, or a cold and wet British spring, as the key obstacles that need to be overcome if they are to survive long enough to pass on their genes.

In writing a book like this, I'm aware that dates may not only vary with location, but also change between years. But as many birds are creatures of habit, with a remarkable sense of timing, most of the key events in the lives of each of the twelve birds can be reasonably closely synchronised across most of Britain, in most years. Also, while trying to piece together the jigsaw pieces as to exactly where and what each of our twelve species are up to each month, the realisation dawned on me that even for our most familiar birds, we still know a lot less about their lives than we think we do. The Blue Tit, for example, tops the 'most intensively ringed bird species' poll every year, but the specifics as to exactly who they associate with, and how far they range during winter, is still far from clear.

My great hope is that this book will encourage and inspire all of us to take another look at what the birds in our back garden and beyond are *actually* doing. Because it's only when enough people take the time to observe, study and enjoy the behaviour of these fascinating birds that we will finally begin to understand why so many continue their alarming decline… and if anybody sees a Nightingale in November, then please let me know *exactly* what it was doing!

Mike Dilger
December 2015

The Species

Bewick's Swan

The first arrival of Bewick's Swans back from their summer break in the Arctic Russian Federation, is always a red-letter day in the life of the few select sites lucky enough to host this spectacular bird in winter. Bewick's Swans are long-lived birds that tend to practise the art of monogamy, with many parents often staying as a pair, quite literally, until 'death do they part'. Taking a long, circuitous migratory route from their breeding grounds as a family, these family units will also play an important part in the swans' winter life too. Those parents returning from a successful breeding season with up to two or three youngsters, who are willing to back them up in any disagreement with other swans, will quickly become a powerful and domineering unit. Certainly at those locations where many swans are vying for the best feeding and roosting spots, pairs without cygnets and single birds will be pushed down the pecking order as family groups bulldoze all before them out of the way. Feeding on anything from aquatic plants to the remains of cereal and root crops in fields, it is vital that as many calories are consumed as possible before the Bewick's Swans attempt the tough return leg back north of the Arctic Circle in the following spring.

Peregrine

The self-styled fastest bird on Earth, the Peregrine has bounced back from an almost terminal decline in its fortunes, caused primarily by the chemical DDT which was responsible for eggshell thinning certainly into the 1960s. As well as occupying their traditional home of mountain crags and sea cliffs, Peregrines have recently begun to enter our urban landscapes, attracted by the high-rise buildings and never-ending supply of pigeons and other city-dwelling birds. Any established pairs will be keen to stay close to their breeding territory for most of the year, which will usually begin with spectacular display flights, before the first eggs are laid in early spring. Any young successfully fledging will be at the bottom of a very steep learning curve as they take tuition from their parents in the art of hunting. Harnessing their speed and killing techniques will be vital for the young birds as they begin to venture out on their own as summer finally turns to autumn.

Lapwing

Also called 'Pee-wit' or 'Green Plover', there can be few British birds with as many alternative names as the Lapwing. With its big butterfly-like wings, there are very few as characteristic in flight either. A classic bird of farmland and marshland, the Lapwing has a breeding distribution across most of Britain and, despite recent declines, around 150,000 pairs still attempt to breed here each spring. Because of Britain's relatively mild winter climate, the resident population can be boosted to as high as a million birds by autumn, as immigrants flood over from the colder climate generally experienced in northern Europe. As the Lapwing is a bird that struggles to feed on frozen ground, on the few occasions when Britain in its entirety is hit by sub-zero temperatures, many of these birds will have little choice other than to simply pack their bags and take themselves off for a spot of French or Spanish winter sun – only to quickly return to more familiar surroundings once the thaw sets in.

Puffin

The Puffin's smart summer plumage, ornate and multi-coloured bill and engaging character put it at the top of the 'must see' list of most novice birders. But away from its remote breeding sites in northern and western Britain, very little is known about the movements of this truly pelagic and mercurial bird. Returning to their colonies in late March or early April, the Puffins will spend the first few days 'rafting' at sea, before finally plucking up the courage to return to terra firma. Pairs often mate for life, and there will be a good deal of spring cleaning of the burrow and cementing of bonds before the single egg is finally laid in late spring. Initially incubated underground by both parents, the chick will then continue its subterranean existence for another six weeks before the strategic withdrawal of food finally sees the youngster forced into action. Emerging at night to scramble down to the sea, the young Puffin will immediately have to fend entirely for itself. By late summer, traditional breeding colonies will once again be Puffin-free zones, as the birds choose to spend the rest of the year out at sea, in a wide arc anywhere from northern Norway and the Faroe Islands, across to Newfoundland and down to the Canaries.

Cuckoo

The Cuckoo's call must rank as one of the most distinctive and iconic calls of any British spring. By manipulating host species like Meadow Pipits and Reed Warblers to carry out the exhausting work of raising their young, once mating and egg laying is completed, there will be little incentive for the adult Cuckoos to spend any more time in Britain before returning abroad. The exact location in which Cuckoos overwinter has been little more than guesswork until recent advances in satellite telemetry revealed that the majority will head for the impenetrable swamp forests of the African Congo. It has also been discovered that on their journey south the Cuckoos will have to choose between entering Africa via Italy or Spain – a case of two routes but one destination! Brought up by foster parents, upon fledging the youngsters must have the African tropics hardwired in as their destination, once they too leave Britain later in the autumn.

Tawny Owl

Heard far more frequently than seen, the classic hoot of this most suburban of all our owls, along with the bark of a Fox, must represent the classic soundtrack to a cold winter's night. Feeding primarily on woodland mammals or birds, many established Tawny Owl pairs will spend their entire lives within carefully marked-out territories, which they will defend both vociferously and physically, particularly during the breeding season. As nesting begins comparatively early in the year, the young Tawny Owls will often fledge well before many summer visitors have even hatched their broods. However, once out of the nest their extended adolescence means they will both be reliant on their parents' support and tolerated under the same roof until at least the autumn. Only once they have grasped the basic fundamentals of fine dining will they be unceremoniously ushered out by their parents, in order to try and find a territory to call their own nearby.

Kingfisher

Spending large amounts of time motionless at the water's edge as it looks for fish, it is perhaps not surprising that few people not keen on birds have ever had good views of a Kingfisher. When not keeping a low profile, this electric blue and orange bullet can be a remarkably common fixture as it whizzes on whirring wings up and down many clean rivers, streams and waterways across much of lowland Britain. The Common Kingfisher is our only native species of kingfisher, and the majority of individuals will be reticent to move from their summer breeding territories as the days begin to shorten and the temperature drops. In fact most Kingfishers will dig in like a tick during the winter, only ever budging when forced out by competition from other birds, heavy floods or freezing conditions. Even for those birds able to find reliable food sources, severe winters can have a devastating impact on the population. However, with the hardy survivors capable of rearing two or three broods of up to six young in their fortified subterranean nests during subsequent breeding seasons, the Kingfisher is also a bird capable of great powers of recovery.

Swallow

Resplendent with its iridescent blue-black plumage, rusty-red face and long tail streamers, the Swallow is surely one of the best known and elegant of all avian visitors to grace our British summer. Clocking up an extraordinary mileage each year, Swallows make the arduous journey to Britain to capitalise on the seasonal yet bountiful supply of insects to rear at least a couple of broods during the long summer days. After moving down towards the south coast by autumn, crossing the English Channel on their southbound journey marks just the beginning of an epic migration, which will entail travelling through Europe and across the Mediterranean, before then flying over the Sahara on their way to southern Africa. Having arrived in their winter quarters, the Swallows will then divide their time between hawking for insects around the legs of elephants and gathering at huge night-time roosts. Only in late winter will they finally begin to leave southern Africa, as they undertake the 9,600km return journey, with the aim of arriving back in the rural British countryside by the following spring.

Robin

Most people consider their own familiar back garden Robin to be a year-long resident. But many human residents of Britain's east coast might be surprised to find 'their Robin' could in fact be an immigrant from the continent, with the local breeding Robin having either opted for a break in another neighbourhood come the autumn, or strayed even further afield for a spot of winter Mediterranean sun. While some Robins exercise wanderlust, many individuals will opt to spend their entire lives within sight of their birthplace, by holding a territory right through the winter. Both a very visible and very vocal presence on their home patch, Robins are able to brighten up the dullest December day with their jaunty song. Already on their breeding territory in early spring, a successful pair should certainly have sufficient time to raise two broods, before the days once again shorten and the available food becomes too scarce to support any more than just one beak.

Nightingale

When it comes to appearances, the Nightingale can hardly be called devastatingly handsome, but then who needs to look like George Clooney when you have a voice with more range than Pavarotti? The Nightingale's rich, melodic and powerful song has been celebrated for generations, but scour the literature and surprisingly little is known about the bird's life history. Famous for skulking, most birdwatchers can count on the fingers of one hand the number of decent views of a Nightingale they've ever had. Arriving in its British breeding grounds in spring, after the most incredible journey which began in West Africa and included both flights around the Sahara and over the Mediterranean, the male must first carve out a territory and find a mate. The main tool the male will use to achieve these twin aims is its beautiful and technical song, which upon arrival will initially be belted out day and night across all southern and eastern counties lucky enough to host this marvellous migrant for a few short spring and summer weeks.

Waxwing

Even in the darkest days of winter there is one bird that can be guaranteed to bring a bit of glitz and glamour to a cold day's birdwatching. With its expressive crest, pinkish-buff plumage, waxy wingtips, and gregarious and confiding nature, there is little not to like about the Waxwing. What adds even more of a frisson to spotting them is that in some years no more than a few hardy souls reach our northern and eastern shores from their breeding grounds across the North Sea and beyond. However, in other years, when a good breeding season is combined with a scarcity of food on the continent, they'll 'irrupt' and head for Britain in their thousands. In these years, it becomes the continent's loss and our gain, as anywhere from Peterhead to Penzance receives a smattering of these gentle marauding flocks. Seemingly as happy to feed in town as in country, any city with street-lined ornamental Rowans, or garden with Pyracanthas and Cotoneasters, will be effectively laying out the welcome mat for this most benign of invaders.

Blue Tit 

Despite being primarily a bird of broad-leaved woodland, the Blue Tit's enterprising and adaptable nature has enabled it to take advantage of the bountiful supplies of food we leave out. Undoubtedly one the most widespread and familiar of all our garden birds, British Blue Tits are considered to be largely resident, with the territorial male courting his female as a prelude to the building of their nest. Raising their single, large clutch in late April or early May, for breeding Blue Tits it really is a case of 'all your eggs in one basket'. Anyone who regularly feeds the birds all year will also be familiar with the sudden influx of yellow-faced juveniles piling in for a free hand-out as broods up and down the land fledge by midsummer. Once outside the breeding season, any territoriality that existed around the nest will disappear, as both adults and juveniles band together with other species to form roving, mobile flocks in search of food. As each species in these mixed flocks occupies a subtly different feeding niche, there should be little direct competition for food, and also more pairs of eyes able to look out for the local Sparrowhawk!

January

As the dust settles on our New Year celebrations, we shouldn't let 'auld acquaintance be forgot' and spare a thought for the welfare of twelve iconic birds that we deign to call British. Winter will still be deepening its icy grip in what is statistically the coldest month of the year in Britain, and having passed the winter solstice only a few weeks previously, each 24-hour period will still consist of close to two-thirds darkness. With most plants lying dormant at this time and a whole host of mammals, reptiles and amphibians choosing hibernation to see out the winter, birds will have one of two choices – to stick or to twist! Despite the cold, dark days predominating, only our summer migrants from our chosen twelve species will have forsaken the British winter for foreign climes. With all three traditional summer visitors bringing in the New Year at very different locations across the African continent, and our British Puffins widely dispersed out at sea, the remaining eight should still be more than able to eke out a living in a frosty Britain.

Early January

Positively revelling in the cold weather, Bewick's Swan numbers will be at their peak here in early January. Having opted to spend the winter over 3,500km from their breeding grounds on the Russian tundra, the Bewick's will be positively basking in the relatively balmy conditions a British winter has to offer. With their summer nesting locations currently covered by snow and ice, ground temperatures plummeting to below −20°C and daylight ferociously truncated, decamping to a northern Europe warmed by the Gulf Stream is frankly the only option for even these hardy swans.

With worries that the western population of Bewick's Swans has been falling for a couple of decades, the 7,000 or so currently wintering in the British Isles is thought to represent close to 40% of the entire European population. The Ouse Washes holds the majority of these British-wintering Bewick's Swans, with over 5,000 spread out at a string of sites, but undoubtedly the best studied Bewick's are those that visit the Wildfowl and Wetlands Trust (WWT) reserve at Slimbridge, Gloucestershire. Attracted in with daily supplies of grain and the promise of a safe night's roost inside a huge encircling electric fence, upwards of 350 wild Bewick's Swans will visit Slimbridge's Rushy Pen during the course of the winter. Uniquely for this species, scientists are able to identify individual birds from the endlessly variable patterns of black and yellow on each swan's bill.

Being able to easily recognise individual Bewick's Swans has enabled the researchers at Slimbridge to reveal much about the lives of these birds that would otherwise have remained largely a mystery. Firstly, many of the swans visiting Slimbridge are regulars, with individuals recorded back in the Rushy Pen over a succession of winters. Also, the Bewick's are capable – with accrued experience and luck – of reaching a ripe old age. A Bewick's Swan called 'Caper', for example, was recorded in the winter of 2014/15, still going strong at the grand old age of at least 26. When not helping themselves to the free handouts from the Slimbridge wardens, the swans will pass much of the time feeding in the abundant grass pastures that dominate estuarine Gloucestershire. Safe in the knowledge that they are both well protected and that disturbance is kept to a minimum, they're able to fill their bellies on a rich grass sward from dawn to dusk, before returning to the sanctuary of Slimbridge each night.

Only ever seen here between October and March, the Bewick's Swans are firmly in the 'winter visitor' category in the minds of British birdwatchers. The Waxwing is the only other bird in our 'top twelve' that does not breed here either, saving its visits to British shores for the shortest days and coldest nights. While no British winter will ever be 'Waxwing-free', the number of these gorgeous 'pink punks' visiting can vary enormously from year to year. In some years, the numbers visiting the UK will be no more than a trickle, but in other winters they seemingly pour over from their breeding grounds in the northern taiga forests of Norway, Sweden, Finland and the Russian Federation. Over the previous 11 years, we've seen three invasions – during the winters of 2004/05, 2010/11 and 2012/13. In these years, up to 10,000 birds were recorded at the height of winter, split into flocks of varying size. Usually arriving along our east coast, and dropping in anywhere from the Shetland Islands to the Isle of Sheppey, flocks of different sizes will descend straight down to the numerous berry-laden trees to refuel after their long flight across the North Sea. Often seen in handsome numbers in our towns and cities, because many local councils have a predilection for lining streets with fruit-bearing trees, the Waxwings will firstly strip this supply before being forced further inland to find more fruit.

Dedicated insectivores during the high summer of the breeding season, by the time winter arrives Waxwings will have made the switch to fully-fledged frugivores, devouring anything from rowan and hawthorn berries, to rosehips and apples. It seems in those 'non-invasion' years, despite the sub-zero temperatures close to their breeding grounds, and days with little more than six hours of light for feeding, the local berry supply should be more than sufficient to keep the vast majority of birds well fed. This means that in the years when only a trickle reaches eastern Britain, there will be far less competition for the berries, and so little need to forage further inland. For reasons that are not entirely understood, in other years, the berry crops close to the Waxwings' breeding grounds may well fail to materialise, and when this is combined with a series of productive breeding seasons the birds will need to migrate rather than risk starvation.

Having crossed the North Sea to arrive in Britain, by early January in an invasion year, Waxwing flocks may well be reported from the West Country, along the south coast and even as far away as Ireland. Being nomadic birds, and driven on by their stomachs, these Scandinavian immigrants can be seen anywhere from out-of-town supermarket car parks to tree-lined suburban streets. Seemingly incredibly confiding, possibly because they rarely come into contact with us humans on their breeding grounds, it is certainly a 'pink-letter day' when a small flock of these enchanting winter visitors pays a visit to a berry-laden tree or bush near you.

Any hungry Waxwings seen dropping into gardens in early January may well at some point be rubbing shoulders with the local Robin holding its territory right through the darkest days of winter. The New Year finds the resident Robin in fine voice as it reminds any neighbouring red-breasts to keep away from his patch. Well aware that its survival depends on keeping this self-proclaimed territory to itself, for the males, the song in early January will also begin to serve the second purpose it was designed for – to attract females for the oncoming breeding season. With the female Robins also having held a winter territory, which they will have defended as if their life depended on it, by now many will have abandoned their winter residence in order to find a suitable breeding partner elsewhere.

The Robin's song has to surely be one of the most characteristic sounds of the British countryside. It should also be one of the first songs for the beginner to learn, primarily because it will be the only bird singing persistently at the turn of the year. It is also one of very few bird species where the female is equally capable of singing as the male. The song is a beautifully melodious warble, and it's hard not to agree with the author Nicholas Cox who wrote in the *Gentleman's Recreation* in 1674: 'it is the opinion of some, that this little King of birds for sweetness of note comes not much short of the Nightingale'.

The turn of the year will also see the local Tawny Owls busily advertising possession of their territory to warn any rivals of thinking twice before trespassing. Unlike the Robin, which in most cases will mate with a different partner each breeding season, an established pair of Tawny Owls is thought to stay together 'till death do them part'. With an intimate knowledge of their territory, on which they will rely entirely for food and accommodation throughout the year, it is vital that any other intruding Tawny Owls are driven off with the breeding season just two months away. The hoot of the Tawny Owl is surely the most familiar call made by any British owl, and is always used, along with the Fox's bark, to signpost 'winter at night' in any fictional drama on television. The male's call begins with the familiar drawn out '*hooo*', followed by a subdued '*hu*' and then a final '*huhuhuhooo*' in vibrato. The female also hoots, but it sounds squeakier and her '*kee-wick!*' contact call is far more characteristic. At this time of the year, the pair will effectively duet as they keep in contact through the hours of darkness, and when the male tags his territorial hoot on to the female's contact call, then the classic '*twit-twoo!*' can be heard.

Traditionally a woodland bird, the Tawny Owl is surprisingly adaptable and will often make a home in urban and suburban areas with mature trees, such as large gardens, parks and cemeteries. The Blue Tit is another one of our chosen twelve with historical woodland associations that has learnt to reap the dividends of living alongside us humans. Each year the RSPB carries out a Big Garden Birdwatch in January, in which the charity persuades over a half a million people to make both a note of the different birds and the abundance of each species coming into their gardens over the space of an hour. With the Blackbird usually topping the poll as the species most commonly recorded, it is little surprise that the Blue Tit is often vying with the Robin for runner-up spot. In 2015, for example, at least one Blue Tit, and probably many more, was recorded in 82% of the gardens surveyed, as the bird's natural acrobatic ability of foraging on the slenderest of twigs is effectively utilised to take advantage of the vast array of foodstuffs that many of us leave out. In addition to large numbers of Blue Tits supplementing their diet from feeders at this time, many of the local birds will have been – since the previous autumn – active and paid-up members of a mixed flock, containing other species such as Great Tits, Goldcrests, Treecreepers and Wrens, amongst others. This gang's remit will be to roam locally for food in the surrounding town and country, secure in the knowledge that this 'safety in numbers' strategy will hugely reduce the chances of each individual being picked off by a hungry Sparrowhawk.

For urban Peregrines, this time of the year should see any established pairs staying close to where they bred the previous summer. On occasion they will even visit the nest-site to check it out for later in the year, but they will also avail themselves of many other buildings during this time. Midwinter is a time when the Peregrines are far less obvious, as most of their time is taken up by roosting and hunting. The warmth of urban areas and an abundant supply of 'flying food' in the form of pigeons may also persuade many rural Peregrines to try the 'city slicker' lifestyle during the winter months too. At this time of year, obvious landmarks might see a number of Peregrines passing through, many of which will be either juveniles wandering well away from where they hatched, and non-breeding adults. These itinerant birds will take advantage of the lack of territoriality of any incumbent pair to undertake fact-finding missions for future breeding seasons, when they too might be in the market for a prime piece of Peregrine real estate.

Also keeping a low profile in early January will be our Kingfishers. Like Robins, the antisocial Kingfisher will only deign to share its territory with a mate during the breeding season, preferring instead a solitary lifestyle for the rest of the year. For the male, his winter domain will commonly be the same patch as his summer breeding territory, with the female either moving away or holding an adjacent winter territory. In equitable circumstances, and with unnerving human parallels, the separated couple may arrange for their previous summer estate to be split down the middle, only to then rejoin forces and territories the following spring. But under the terms of their winter separation, there will invariably be a strongly delineated boundary, across which trespassers will most definitely not be welcome.

Apart from the odd territorial dispute, the only issues to preoccupy our Kingfishers as the New Year begins are if high winter rainfall causes flooding, or if a late freeze makes fishing difficult. Very cold winters can have a devastating impact on Kingfisher populations. If their feeding areas freeze over, hungry birds must move to avoid starvation or find different food. The winter of 1939/40, for example, was so severe that the River Thames completely iced over for a time, causing the Kingfisher population to plummet from around 120 birds along a 68-mile stretch, to just a couple of pairs once the river had finally thawed.

Another of our 'twelve avian disciples' that fares poorly during cold snaps is the Lapwing. As an invertebrate specialist throughout the year, a covering of snow or hard frost means that the Lapwing will either be unable to uncover sufficient food or winkle it out from the frozen fields. Many Lapwings that breed in the UK will also winter here, forming large, mobile flocks that scour the countryside looking for feeding opportunities. Boosted by huge numbers of continental Lapwings, which have been pushed across the North Sea into Britain by 'Baltic' conditions elsewhere, the Lapwing population may be as high as 620,000 individuals by the height of winter. With a breeding population of around 130,000 pairs currently in Britain, this gives some idea of the huge continental influx of this fair-weather wader.

Plummeting temperatures are of course not an issue for the Nightingale, which will have passed through the Iberian Peninsula during August and September of the previous year, on the way to spending a winter in West Africa. A secretive bird at the best of times, where the British Nightingales spend their winter was unclear for a long time. According to the British Trust for Ornithology (BTO) a paltry total of just two Nightingales originally ringed in Britain have subsequently been re-trapped south of the Sahara – hardly a sound dataset to give a clear picture of their movements. However, this is now all changing thanks to the use of tiny 'geolocators', which have recently been attached to a number of Nightingales trapped in mist nets on their British breeding grounds.

No larger than a shirt button, and weighing just 1g, geolocators are revolutionising the way small birds can be followed on their migration. The devices have an inbuilt clock, calendar and a light sensor, and constantly monitor the daylight against the time and date. Attached to the lower back of the bird, and held in place by two loops running around the bird's legs, once the geolocator is recovered from a returning bird, the researchers should then be able to calculate where on the planet the geolocator, and of course the bird, was at any given time and date. In 2009 the BTO attached geolocators to male Nightingales on their breeding grounds. Nightingales show huge fidelity to their breeding sites, and so the following spring six of the geolocators were recovered from returning birds and analysed. Unfortunately five failed to produce any meaningful data, but the now famous Nightingale OAD (so named by the letters on his device) was successfully tracked all the way to sub-Saharan Africa before failing.

Dr Chris Hewson, the lead researcher, has said that 'we have learnt more from this one bird – OAD – than in the previous 100 years of ringing Nightingales!' The upshot of all this cutting-edge technology is that we now believe that most of the Nightingales which come to breed in England will be in the West African countries of Sierra Leone, Guinea or Guinea-Bissau come early January. The BTO has also led the way in the satellite tracking of Cuckoos in recent years, which is finally revealing the mystery of where British breeding Cuckoos go when they leave our shores. Transmitters attached to birds, which are then tracked by satellites, have recently been miniaturised to such an extent that they can now be placed on birds as light as a Cuckoo. Unlike the geolocators, which need the individual birds to be recaptured to retrieve the data, for the last four years Cuckoos have been tracked along both their migration route and into their wintering grounds in real time.

With just one previous record of a British-ringed Cuckoo recovered south of the Sahara, in Cameroon in (believe it or not) 1930, the satellite data is finally uncovering what was one of our great ornithological mysteries. Fifty-four Cuckoos have been tagged over the duration of the project, but only one individual managed to survive the research project's first three years with its 'on-board' transmitter functioning properly throughout. Chris the Cuckoo, named after the TV naturalist Chris Packham, before the bird's untimely demise in 2015 successfully navigated the perils of the Sahara Desert seven times, collecting ground-breaking data every step of the way! The information collected from Chris, and other male Cuckoos (as the females are still considered too light to carry the current model of transmitters), has so far uncovered quite a number of revelations. Come the turn of the year it seems most of the British-breeding Cuckoos are still firmly settled in one of two wintering quarters, either the swamp forests of the Republic of Congo, close to the border with the 'other bigger Congo', the Democratic Republic of Congo, or further south in the altogether more arid habitat of northern Angola.

Remarkably the Cuckoos are still not the most southerly of our twelve birds, with British Swallows having flown down to winter some 2,800km even further south. For centuries, after Barn Swallows disappeared from Britain in the autumn, they were believed to spend the winter in rock or tree crevices, or even buried in mud. It was not until the early nineteenth century that observations of birds over the English Channel, in the Mediterranean and off West Africa, supported the alternative, and at the time controversial, view that they migrated. Only in 1912 was definitive confirmation of the British Swallows' winter home produced, when a female Swallow ringed in Staffordshire by a James Masefield in May 1911 was recovered in Natal on 23 December 1912.

From the BTO's ringing records, it seems that most British Barn Swallows winter in South Africa, with a large number of records predominating from the Western Cape Province, an astonishing distance of around 9,600km from south-east England. Roosting by night in reedbeds or other wetland vegetation, sunrises in early January, representing high summer in the Western Cape, will see the Swallows leaving in waves to fan across the South African countryside to feed. In Britain, the airspace around cattle and horses is often a productive feeding ground as their dung attracts insects and the livestock also flush invertebrates from the vegetation as they walk. In South Africa, herds of game may well offer similar feeding opportunities. A British Swallow swooping to catch a dung fly from beneath an elephant's legs?… not as far-fetched as it sounds!

Our final iconic bird, the Puffin, may well be somewhat more difficult to track down in early January. Having left their breeding colonies by mid-August at the latest, current research by Mike Harris and his team working on the Isle of May, Firth of Forth, eastern Scotland suggests that most of their Puffins immediately move out into the North Sea. Using geolocators attached to the Puffins, the team found that after then flying around north Scotland to spend the autumn and early winter in the north-east Atlantic, many will have already returned to the North Sea by the turn of the year. Recoveries of Puffins from colonies elsewhere in Britain suggest they disperse widely, with recoveries of birds seeing them turn up from Brittany in France to the Canaries, and a few adventurous juveniles even making it as far as Newfoundland and Greenland.

Mid-January

With Bewick's Swans firmly settled on their wintering grounds for more than a couple of months, and dominance hierarchies mostly established, aggressive encounters between birds should now be far fewer than earlier in winter. Top of the 'pecking order' will be the families, consisting of a pair of experienced adults and their juveniles which were successfully raised during the previous breeding season. Sticking close together at all times, a family of two adults and three offspring represents a formidable fighting force that is easily able to bully those swans with fewer allies away from the best feeding locations and roosting sites.

Nowhere is this hierarchy more noticeable than at the WWT Slimbridge reserve in Gloucestershire. With the swans fed daily by the wardens, the families will use their superior numbers to barge their way to the front of the queue, and help themselves to far more than their fair share of the grain being handed out. Aware of the fact that numbers equate to dominance, yearlings (swans experiencing their second winter), will often rejoin their parents on the wintering grounds to boost both their own social position and that of their parents within the flock. Any trios consisting of parents and a yearling, although often not able to compete with larger families, will still be able to exert influence over those pairs of swans without any young at all to back them up. Below the established pairs without any young, and even lower down the pecking order, are those adults that have either lost a mate or have been unsuccessful in managing to secure one. And finally, the birds that will struggle the most during feeding and roosting will be those yearlings battling the unfortunate combination of a solitary existence and huge inexperience.

Compared to the highly strung Bewick's Swans, there generally seems far less aggression amongst the socially gregarious Waxwings. By mid-January in a good 'Waxwing winter', the overwintering population may by now have fanned right across the southern part of the UK, with some hungry birds even braving a second sea crossing to sample the delights of Ireland's berries. Staying in flocks is initially an asset to Waxwings, as it enables them to use their numbers to dominate prime feeding positions, in the process keeping territorial Mistle Thrushes, for example, away from the berries. However, as the fruit resource becomes far more depleted at this time of year, any larger flocks may well be forced to break up into smaller, nomadic groups. These compact, mobile flocks will then need to range far and wide to ensure they can find enough food to see them through those nights when the temperature dips below zero.

For many of our resident birds that have chosen to stay at home rather than taking a long winter break in warmer climes, extended cold periods can present a very real and imminent threat to their survival. Both Kingfisher and Lapwing from our chosen twelve will struggle to find enough food if the water and soil surface stay frozen for any length of time. According to the BTO, as many as four out of every five Kingfishers that fledge won't see even their first birthday, with a large proportion perishing during periods of severe weather. Struggling to find unoccupied riverine territories, many young birds are pushed in the autumn to spend the winter at the coast to feed in estuaries or even in the sea itself. Although these habitats will rarely, if ever, freeze, the feeding around the coast will always be far more demanding than, say, a quiet chalk stream in Hampshire.

In very harsh conditions, some Kingfishers holding inland territories may be forced to temporarily move if their established feeding areas freeze over. The cold weather movements of continental Kingfishers are well known, and particularly in Germany, where the Kingfisher's name 'Eisvögel' translates into 'ice bird'. Perhaps a technically more accurate name, although undeniably more of a mouthful, would be 'the bird that moves ahead of the ice'!

After heavy snow, or during periods of hard frosts, many Lapwings from northern Britain tend to be either pushed towards south-west Britain or across to southern Ireland, which tends to avoid the worst excesses of winter, thanks to the warming influence of the Gulf Stream. This contrasts with the many Lapwings that have bred in the southern half of Britain, which seem to respond to freezing conditions by simply abandoning the British Isles entirely. Crossing the English Channel, most of these birds will temporarily settle anywhere from coastal France all the way down to the Atlantic coasts of Spain and Portugal, before the 'breeding imperative' draws them back to Britain later in the spring.

One bird that will refuse to budge from its territory, come rain, shine or even Arctic conditions, is the Tawny Owl. An established pair will rely on an intimate knowledge of their territory to eke out an existence even in the most severe of conditions. For those 'Tawnies' that have set up home in more urban and suburban environments, recent research suggests that they take a higher proportion of birds in their diet, meaning their choice of food may well be easier to find after a covering of snow. This contrasts with many rural Tawnies, which mostly rely on small mammals for sustenance throughout the year, with Bank Vole and Wood Mouse being the two species most commonly encountered in the owls' pellets. Unlike many British mammals, the Bank Vole and Wood Mouse do not hibernate, but stay active all winter, with the Wood Mouse venturing out to forage mostly at night, while the Bank Vole will feed intermittently right around the clock. Snowy conditions, for example, can make hunting the owls' prey far more difficult, as the mice and voles will still be going about their business, but out of sight under the snow. In this situation, the Tawny Owls may supplement their diet with carrion from animals that have succumbed to wintry conditions. This was clearly demonstrated in the BBC's *Winterwatch* in January 2015, where the team managed to film a Tawny Owl tucking into a deer carcass each night at the Mar Lodge Estate, within the Cairngorms National Park. Tawny Owls, it seems, are not just top predators, but supreme opportunists as well.

Harsh British winters can exact a toll on even the hardiest and most resilient of birds, such as Robins. With insects providing the bulk of its diet for most of the year, the Robin's ability to change to fruit and seeds during winter stands it in far better stead than birds such as Treecreepers and Dartford Warblers, which are confined to an insectivorous diet throughout the year. It will not just be the Robin's catholic diet that helps see it through any particularly cold periods, but also its willingness to take advantage of our love of feeding garden birds. Being so confiding by nature has enabled the resident Robin to often be first in the queue when it comes to dining out on any food we leave out on bird tables up and down the land.

Unlike the Robin, the holding of a territory is not an issue for Blue Tits in the winter, resulting in large numbers of individuals continuing to move through our gardens at this time of year. Frequently accompanied by other small birds, the majority of Blue Tits passing through will be local birds that have bred in neighbouring gardens or nearby woods, with a varying number of nomadic birds added to the mix. Most British Blue Tits are not thought to move more than a few kilometres from where they hatched, but by contrast continental Blue Tits are much more inclined to migrate in winter, even crossing the North Sea in some years. So for those with gardens on the east coast, there is every possibility that some of the birds seen feeding on peanuts or fat balls could be Blue Tits of Belgian, Dutch, German, or even Norwegian descent.

While many Peregrines breeding in northern Europe, North America and Siberia will be forced to migrate south to find enough food to hunt, our comparatively mild winter climate means most British Peregrines will rarely have to move far from their nest site even when the temperatures drop. Although some upland Peregrines will move away from their breeding quarters as winter sets in, most urban pairs should be able to find enough feathered prey in and around their towns and cities to sustain them through to spring. Recent research carried out by Peregrine expert Ed Drewitt suggests that urban Peregrines might not just be hunting during the day, but are making a substantial number of kills through the night too. By examining prey remains, mostly in the form of feathers collected at well-established plucking sites, Ed has been able to reveal that Peregrines are taking a far wider range of food than just feral pigeons. Winter visitors to Britain such as Woodcock, Teal and Moorhen are also being taken at this time of year as they migrate across urban areas at night. Historically, these birds may well have formed a much smaller part of the Peregrine's diet; however, it is thought that as the light from street lamps and buildings shines into the night sky, this light pollution is illuminating them as they pass overhead.

While Peregrines are most commonly seen catching birds during the day with the stoop dive, which involves closing their wings and dropping out of the air at immense speeds to hit and hopefully kill outright their chosen quarry, at night they employ a totally different technique. Lurking in the shadows of a church or tower block, when an urban Peregrine spots the pale belly of a bird lit up like a light bulb, it only need fly a short distance upward to take the bird unawares from below. Peregrines will also insure against any potential temporary shortages by storing excess food during times of plenty. In winter this means that any cached food will effectively be stored in the equivalent of a fridge freezer!

British-breeding Puffins, come the middle of January, will still be dispersed in a wide arc between the North Sea and the north-east Atlantic. As many birds are already in winter plumage by the end of August, they are thought to undergo a complete moult between late autumn and early spring to ensure they look freshly laundered by the time they arrive back at their breeding colonies the following spring. It's not clear exactly when the main flight feathers are replaced during the winter, but recoveries of dead Puffins either hunted or washed up ashore during this time seem to show that the moult is carried out in a synchronous fashion along each wing. While these feathers are growing the Puffins will become flightless, but despite an inability to take to the air, they are still able to dive and search for food. Presumably any Puffin still renewing its flight feathers at this stage will need to have made sure its chosen moult location will be able to provide enough food until its wings are finally ready for taking to the air as well as cutting through water.

Unlike the Puffin, which will be bobbing around out at sea like a cork for at least four to five weeks while moulting its flight feathers, our British Swallows down in South Africa will still be able to fly throughout their entire moult. Because Swallows obviously need to be able to fly efficiently in order to feed, the moult is a slow process which may take anywhere between four and a half and six and a half months.

The Swallow's body and head feathers are the first to moult, followed by those on the wings and tail. They must also moult the flight feathers in a symmetrical, ordered fashion along each wing to prevent any imbalance during flight and ensure their manoeuvrability is not impaired while feeding. With these feathers being replaced so gradually, it is quite normal for the outer primary and tail feathers to still be actively growing by the time the birds have begun their long spring migration back to Britain in around a month's time.

Meanwhile, further north it seems from telemetry data provided by the BTO that most British-breeding Cuckoos will be in a broad belt stretching along the mighty Congo River in mid-January. The river also marks the boundary between the two Congo countries – the Republic of Congo to the west of the river and the Democratic Republic of Congo to the east. This border region is dominated by Western Congolian swamp forest which, when combined with the Eastern Congolian swamp forest, is thought to be one of the largest continuous freshwater swamp forests in the world. Known for its large numbers of Lowland Gorillas, the Western Congolian swamp forest consists of seasonally flooded forest with a high canopy, dense undergrowth and a muddy floor, which is also interspersed with grassland areas. Lush Raffia palms and low vegetation, mixed in with swamp and abundant waterholes, makes passage through this largely pristine habitat almost impossible for man.

Due to the more imprecise positioning of geolocators, when compared to the Cuckoos' satellite transmitters, it's difficult to pinpoint which habitat Nightingale OAD was using during its stay in Guinea. Since the original pioneering work tracking

OAD, five more British Nightingales have been successfully tracked all the way to West Africa, with most seeming to spend mid-January along the coastal strip between south-west Guinea and north-west Sierra Leone. Having spent a month up to mid-December in coastal scrub further north in Senegal and The Gambia (Senegambia), Chris Hewson from the BTO thinks that as Senegambia becomes too dry at the end of the year, the Nightingales may well be moving south up a 'moisture gradient' to an area where the rainy season has only just finished, potentially making it easier for the birds to find food. In other words, with the start of their return journey to England just two months away, moving even further in the other direction seems to be the smartest move!

Late January

Leading the headlong charge towards the breeding season will be the Tawny Owls. Any established pair will have already been holding exclusive access to a territory throughout the winter, and the next logical step will be to select the precise nesting location in which their clutch will be laid and resultant brood reared. The end of January will see the parents visiting all the potential sites within their territory before deciding which suits their precise requirements best. The Tawny Owl is of course primarily a cavity-nesting species, often favouring large holes in deciduous trees, although they will sometimes resort to old stick nests or even sites in buildings if push comes to shove. In territories that consist mostly of coniferous woodland, with few natural cavities, they may even nest in amongst rocks or tree roots on the ground. More recently, where natural options may be scarcer, many Tawny Owls have been persuaded to use nest boxes, with a considerable degree of success. The recent advances in miniature camera technology have meant that for any box-nesting Tawnies we're now in a position to learn much more about their secretive nesting behaviour.

Certainly in southern Britain, most Robin pairs will already have become established by the end of January. Pair formation is considered distinct from courtship, which is usually nothing more than a brief precursor to when nesting begins properly in the spring. Like most species of bird (and indeed most wildlife), it is ultimately the female Robin who will choose the male. So with the males already singing lustily at this time of year, the onus will be on the bachelors to entice any unpaired females into their territory. With a surplus of unmated males each year, the female can afford to be quite fussy, and will sometimes move into the territory of a single male, only to quickly reject him for the neighbouring male if she feels he has more to offer. It's difficult to elucidate exactly how the female makes her choice, but it may well be a combination of the standard of his song, the quality of his territory, his looks and aggression.

A whole neighbourhood of Robins in Devon was studied closely by David Lack for his celebrated book *The Life of the Robin*, and it was David who discovered that Robins commonly form into pairs soon after first light. Pairing up begins with the prospective female flying into the male's territory, and after belting out a brief burst of song she will then fly right up to the male. The Robin's red breast is of course its most conspicuous feature and is frequently used for display purposes, to both impress and intimidate. Initially confronted with an intruding Robin at such close quarters, the male will aggressively display at her, which involves extending himself to show her as much of his bright red breast as possible. With wings flicked half-open, his tail cocked and the feathers on his crown standing upright, he will then often turn slowly from side to side, displaying his breast in a peculiar mechanical motion, which looks distinctly at odds with what is normally a busy little bird. Also replying with song of his own, at some point the penny must drop for the male when he realises that the Robin constantly perching uncomfortably close to him, despite his threat display, happens not only to be female, but also (crucially) a single female. Quickly changing his tune, he then invites her to begin feeding alongside him on the ground.

With most Robin pairs keen to rear at least two broods during the course of the spring and summer, for the single-brooded Blue Tits there is no such hurry. Despite nest-building still being over two months away, it's thought that pair bonds may well have already formed during the winter months, while the tits were travelling around in their mixed flocks. As the pair waits for the longer and warmer days of spring to cement their bonds, the female will already have a clear idea where she'd like to rear her young, and so at this time may begin spending the long winter nights at the exact location she intends to be her future nest site.

As January draws to a close, established pairs of Peregrines will use this time to strengthen their pair bonds close to the nest site. Having tolerated the presence of itinerant Peregrines throughout the winter, a change of behaviour will start seeing them adopt a zero tolerance approach to any intruders violating the pair's airspace. Any Peregrine still unwilling to leave at this stage risks physical violence from the incumbent pair, and trespassing Peregrines have even been killed. Peregrine expert Ed Drewitt reported a dead male found below an active nest in an urban site in Derbyshire in 2013, which was thought to have been killed by the resident pair.

Despite most Peregrines not forming clutches until at least mid-March, many pairs will have already begun copulating – behaviour that may well continue right the way through until the female is finally ready to lay.

Still keeping to their respective territories, our resident Kingfishers will be keeping a low profile along their favoured watercourse in late January, where they'll divide their time between fishing, roosting and preening. Having begun a complete moult of all their feathers after the rigours of the previous breeding season, most Kingfishers spotted tucked in tight against a river bank should by now be looking in pristine condition. Needing to find enough food just for one, they'll nevertheless be surviving on minimal rations as so many freshwater fish will be tucked away and moving little during the winter months.

With any cold weather continuing in the same vein, those British-breeding Lapwings that escaped the worst of the British winter by hopping over to the continent will probably stay away until closer to the breeding season. For those hardier Lapwings that have either stayed in Britain or are cold weather visitors from the continent, late January will still find them crowding together in large, mobile flocks as they take advantage of whatever feeding opportunities present themselves.

During winter, the Lapwings seem to favour either large areas of mixed and arable farming or improved permanent pasture. Where arable land is more common in central and eastern England, Lapwings are often found in the highest concentrations on winter cereals, but they're quick to move to pastures further west if temperatures continue to drop. It seems that pasture is better at insulating the ground, so the

Lapwings may still be able to dig for earthworms when conditions are freezing elsewhere. Moving to coastal farmland is also an option for any Lapwings really struggling at this time of year, as soil temperatures here tend to be marginally higher than further inland. This is because the sea cools down more slowly than the land in winter, and is the main reason why snow on the coast scarcely settles when compared to inland locations at a similar latitude and altitude.

Historically Bewick's Swans coming to Britain for the winter fed on aquatic and marshland plants in wetland habitats such as flooded pasture. But as large tracts of lowland Britain have been drained and turned over for cultivation to arable land, over the last 40 years many of the swans have switched to feed on root and cereal crops. Certainly in Britain, crops such as sugar beet, winter cereals and potatoes have now become prime feeding areas for large numbers of Bewick's Swans, particularly in eastern England. With the Ouse and Nene Washes thought to hold as many as two-thirds of all overwintering Bewick's Swans in Britain, numbers will still be very high by the end of January. Feeding in fields by day, their roost sites will constantly change during the season, according to the water levels on the Washes. By contrast, the diet of Bewick's Swans that have traditionally settled in more westerly areas of Britain will mostly consist of food from the managed grassland, saltmarsh and flood pasture. For the regular swans at WWT Slimbridge, dusk will see them flying in to roost on the protected freshwater pools with bellies full of grass to digest.

Even in those years when relatively few Waxwings have ventured across the North Sea, the end of January will still see a smattering of records, mainly along the east coast. In a big invasion year, such as the winter of 2012/13 for example, late January recorded Waxwing flocks as widely dispersed as west Cornwall, Anglesey in North Wales and even the Outer Hebrides off the west coast of Scotland. It seems Rowan, Hawthorn and Cotoneaster may well be favoured by the Waxwings early in the invasion, but they must also compete for these berries with the resident Blackbirds, Song and Mistle Thrushes and other winter visitors such as Fieldfares and Redwings. This means that as their favourite foods become stripped, the Waxwings have to turn to seemingly less favoured food sources, such as Whitebeam, rosehips, Guelder Rose, Crab Apples, domestic apples, Privet and Mistletoe.

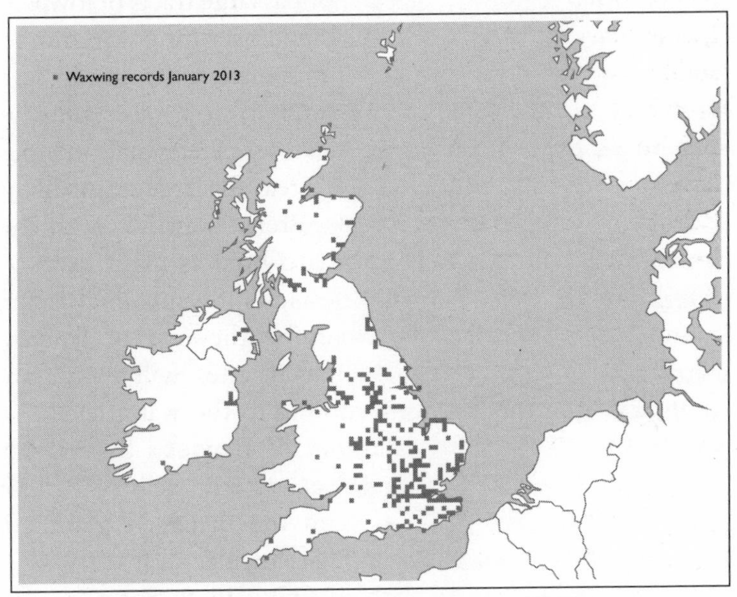

BirdTrack Records of Waxwings reported in Britain and Ireland in January 2013.

With the location that British Cuckoos spend the winter being only recently revealed, information as to what they might actually be eating in the Western Congolian swamp forests or lowland forests of northern Angola is non-existent. In Britain, Cuckoos eat hairy caterpillars, beetles, flies and ants, as well as the females predating eggs and chicks from any nests that they parasitise, and there is no reason to suggest that their diet will deviate dramatically while in Africa. Certainly the forests of the Congo are incredibly biologically diverse, so there must be plenty of invertebrate food on offer for the Cuckoos.

Similarly for Nightingales at this time of the year, there are still far more questions than answers. It's likely that most British-breeding birds are still in the coastal zone between Guinea and Sierra Leone and frequenting savanna woodland, thorny scrub, river gallery forest, humid forest edges and any areas of low secondary growth. Being a ground and perching bird, very fond of keeping to cover and rarely flying out in the open any more than is necessary, it is a safe assumption that they will also be keeping a low profile in West Africa. Before leaving Britain, Nightingales will supplement their insectivorous diet with berries, but it's more likely that on their wintering grounds they will be feeding on beetles, ants and flies caught on the ground and in the undergrowth.

From the little information we have about Puffins feeding away from the colony, the fact that large flocks are virtually never seen out to sea in the winter months suggests that they largely shun company, and being so spread out means they will be feeding at very low densities. Those Puffins that haven't finished their wing moult will still have only a very limited ability to disperse any distance, and it's a fair assumption that when not roosting on the sea, the birds will divide their time between diving for food and preening at the surface. During the breeding season the Lesser Sandeel tends to dominate the Puffins' diet, but during winter, the fish bury themselves into the sand on the sea floor, instantly making themselves more difficult to catch. This in turn means that the Puffins will be forced to turn to a far more varied diet during winter, consisting of other marine fish, marine worms and even squid.

Because of the far more open nature and visible way in which Swallows go about their feeding business, more is known about their winter diet in southern Africa than for any of our other three summer visitors. It seems that in winter, the Swallows' main prey is flying ants and beetles, but they also take spiders, caddisflies and grasshoppers. They will also take non-flying prey like bugs and caterpillars, with swarming termites thought to be an important part of their diet following rains. While flies are a very important part of a chick's diet in Britain, it seems this prey is less frequently caught in South Africa. Being far more mobile, flies are more difficult to catch, and as the Swallows will still be moulting their tail feathers and outermost primaries at this time, they won't have quite the fine degree of manoeuvrability that they would normally expect, so opt instead for the 'easy meat' of slower-moving prey.

February

The weather in February could best be described as 'a curate's egg', or 'good in parts'. On a cloudless sunny day it can feel like spring is bursting forth, only for the following day to bring the harsh reality of winter crashing back down with snow, ice and a brutal northerly wind. With the weather effectively operating like meteorological roulette, the cold, wet and windy days will see our resident birds hunkering down and committing to little more than finding sufficient food to see them through the nights. Better days, however, will kick start the mating game.

Tradition says that Valentine's Day on 14 February is supposed to be the date when our resident birds pair up, and indeed this certainly seems to be the case for birds like the Kingfisher and Lapwing. But many of our resident species, such as Robins, Blue Tits, Peregrines and Tawny Owls, will have already found their partner well before this romantic date and are by now looking to take their relationships on to the next level. Irrespective of their state of readiness, any of our year-round residents will still be further along the breeding process compared to our summer and winter visitors, which will only just be ambling up to the starting blocks in the great race to mate.

Early February

Having secured their territory for the oncoming breeding season, an established pair of Peregrines will use the first fine days of February to strengthen their pair bonds. For Peregrine watchers this represents the start of an exciting few weeks as the pair undertake spectacular courtship manoeuvres on the wing. Co-operative hunting between the male and female can figure highly in these early courting activities. Joining forces to hunt the same bird, the pair will often coordinate their attacks to counter any evasion attempts by their target. By swooping at a flock, in what may amount to nothing more than a dummy run, the male's job will be to try and part an individual from the fold, which then acts as a cue for the larger and more powerful female to stoop at the bird using the element of surprise. Any prey caught successfully in this way will then usually be eaten by the female to ensure her body is in as good a condition as possible for the physical rigours of egg laying ahead.

With the mixed feeding flocks slowly beginning to disintegrate, the first signs of territorial activity amongst Blue Tits will now become apparent on fine days. The winter can prove a savage time for Blue Tits, with only 50% of adults making it through to breed from one year to the next. For the juveniles, of course, it has been an even steeper learning curve since they fledged the previous summer, with a whopping 90% succumbing along the way to starvation, disease or predation. This would suggest that the average pair of Blue Tits may well consist of one experienced bird and another attempting to breed for the first time.

Fine days in early February represent the first time that the male Blue Tits will endeavour to lay down a territorial marker by cranking up their singing from a series of prominent perches. Although not quite hitting the complexity or beauty of either the Robin or Nightingale's overtures it is nevertheless a pleasing little song. Sounding like '*see, see, see-chu chu chu*', with the latter section ending in a short trill, this song will be heard increasingly frequently as the month wears on.

Those Robins heard singing ardently in early February will mostly consist of as yet unpaired males or birds still sorting out territorial issues. Males that have already managed to bag a female by this stage will now begin to sing in a much more subdued manner. Their first few days as a couple will be spent with the male following his new mate around as she explores their now jointly owned territory. It seems she has to learn the boundaries by trial and error, with any incursions into adjacent territories resulting in her being swiftly rebuffed by her new neighbours.

By contrast, having held firm all winter, and possibly even for a number of years, any established pair of Tawny Owls will already be very familiar with the size and shape of their territory. From a range of studies carried out across Britain it seems a Tawny Owl pair's territory can cover anywhere between 12 and 70 hectares, with dimensions depending largely on the quality of the habitat, food availability and density of other Tawnies in the area. The choice of a precise nesting spot within the territory may be a formality if one particular site has been used successfully the previous year. Alternatively, they may attempt to ring the changes following a previous nesting failure. Preparations in the nesting chamber are none too elaborate and amount to little more than the female excavating a shallow scrape that will ultimately hold her clutch, in little more than a month.

Finally shaking themselves out of their winter torpor, early February will see our Kingfishers turning their attentions for the first time to the busy summer ahead. The minority of Kingfishers which held summer territories in more northerly or upland sites, but spent the winter in more benign lowland or coastal locations, may well now be beginning to return to their breeding grounds. Certainly the majority of males will have been holed up on their breeding territories all winter and so have no need to move, but many females will now have to up sticks in order to track down a mate. Kingfishers are mostly monogamous, but won't necessarily pair with the same partner from year to year. It does seem, however, that in those cases when a pair does reunite, it tends to be in those locations where their previous joint summer territory had been cleaved into two winter territories.

While still not quite ready to return to their breeding locations, as the time for departure rapidly approaches, Lapwing flocks will suddenly become far more restless. Bickering between birds can often be seen around this time and sudden erratic flights by birds around the flock can introduce a certain collective skittishness. When birds chase each other on the ground, they will do so with their wings raised to show their striking white wing-linings and use their broad black breast-band to both impress and intimidate. These actions are all dress rehearsals for the territorial and courtship behaviour that will soon be used in earnest on their breeding grounds.

Before departure, from early February onwards, the Lapwings will also undergo a partial moult. Having already completed a full moult between late May and September of the previous year, to insulate them from the worst of the winter, this pre-breeding replacement of feathers on the head, neck and upper breast has an entirely different function – to look good. The change is most marked in the males, as they now develop their characteristic long, wispy crests, boldly marked black and white faces and striking black breast-bands.

With summer arriving in Arctic Russia and northern Scandinavia much later than at the temperate latitudes of Britain, our two winter visitors are in no immediate hurry to leave our shores for their still frozen breeding grounds. Numbers of Bewick's Swans are still high in early February, and as winter draws to a close, those populations that have spent most of the winter feeding on arable crops may well be forced over to pastures as their first choice becomes far more depleted. Also faced with the local exhaustion of food supplies, Waxwings at this stage of the winter will have to range far and wide to get their berry fix.

Where they do manage to locate plentiful supplies of food, the Waxwing's technique often seems to involve little more than eating large volumes of fruit in one sitting. The record for 'prodigious Waxwing eating' must go to a bird seen in Pembrokeshire during the winter of 1949/50. When watched for most of the day observers estimated one individual bird to have eaten between 600 and 1,000 fruits of *Cotoneaster horizontalis.* Unlike many perching birds, Waxwings don't possess a crop in which to store these huge meals, but do have a section of the oesophagus that extends under the skin of the hind neck, which during intense feeding bursts fills in a similar way to a hamster's pouches. Suffice to say that the 90g of fruit estimated to have been eaten by this bird still amounted to twice its own body weight!

Faced with the cold conditions in the North Sea and North Atlantic it's a fair assumption that our British-breeding Puffins will currently be preoccupied with finding enough food to eat throughout the daylight hours, particularly with an energetically expensive moult to be completed. From time-depth recorders placed on Puffins by researchers on the Isle of May during the breeding season, it was found that birds feeding chicks averaged over 1,000 dives per day, spending close to eight hours under water! Despite not having to fly to and from the breeding colonies, as the sea temperatures will be lower and food more difficult to find at this time of year it's a fair assumption that certainly during the short days, the overwintering Puffins will have little time for loafing at the surface.

With the Swallows travelling the furthest back to Britain of our three summer visitors from Africa, it's no surprise that they are also the first to peel away from their wintering grounds. With males invariably arriving in Britain as much as a week earlier than the females, early February will see the very first males begin their migration, but it won't be until later this month that the majority of birds will leave. Roosting communally by night in reedbeds or other wetland vegetation, some sites in South Africa can contain huge numbers of birds.

One particularly well-known location for roosting Swallows are the reedbeds at Mount Moreland in the province of KwaZulu-Natal in eastern South Africa. Despite the fact that most of our Swallows are considered to winter further west in the Western Cape, there will doubtless be British birds mixed in amongst the flock of up to three million that fly in half an hour before sunset after a day's feeding in the surrounding countryside.

From the BTO's work satellite-tracking Cuckoos it seems the majority of British-breeding birds in early February are still in their regular winter feeding grounds at a variety of locations centring on the Congo River. With no more than a few weeks before they begin their long and winding route back to Britain, it's important that they use this time to feed well and ensure they're in the best possible condition for the huge journey ahead. Likewise, our British-breeding Nightingales will still be ensconced in their winter quarters along coastal regions anywhere between Guinea and Sierra Leone. During the winter the males are thought to occupy small territories until they finally leave in mid-March, quickly returning to Britain in a series of very long hops.

Mid-February

With a typical lifespan of six years, and the oldest known bird reaching the grand old age of 17, an established pair of

Peregrines may well be able to hold on to a good territory for at least three or four breeding seasons. Nevertheless, an intimate knowledge of one another will not preclude them from reaffirming their bonds in the most spectacular of ways each February. With last month's co-operative hunts served up as little more than an hors d'oeuvre, the main event will begin to see the pair conducting high-speed chases. Stooping at one another with increasing enthusiasm, the pair can build up to a truly dazzling display. For any town or city hosting a pair of urban-nesting Peregrines, a fine day in the middle of February may well see the birds tearing across the cityscape, in behaviour that is only slightly modified from the basic techniques used for hunting and territorial aggression.

While perhaps not quite as flamboyant as the Peregrine's way of declaring ownership, the male Blue Tit's song is an equally clear statement of intent to any other males in the neighbourhood. On fine days in mid-February a male Blue Tit holding territory will spend a large portion of daylight hours endlessly repeating his simple song to all willing to listen. His song will only be heard less frequently once the boundaries between all his neighbours have become clearly demarcated. Additionally, it's thought each male will come to quickly recognise the distinct songs of their respective neighbours. This means that if a newcomer attempts to barge in by suddenly adding his voice to the chorus, he will instantly stand out. With this unknown male potentially upsetting the stability of a settled community, all the surrounding Blue Tits will immediately take to their perches to send the message loud and clear that the intrusion is an unwelcome one. Any pair without a nesting location nailed down by this stage will begin prospecting potential nest sites. So, from this time, any Blue Tits spotted popping in and out of various garden nest boxes will be obviously shopping around in a 'try before you buy' policy.

By now most Robins should have secured both a territory and a partner, and with these two essential prerequisites in place, it might seem that they would be keen to move on to nest building, but this doesn't seem the case. Stalling any progress, the pair then enter an extended 'engagement period' for a few weeks. During this time, the male and female seem to virtually ignore each other and, despite staying in the same territory, are seldom seen together. The female at this point tends to become shy and retiring and the male takes to singing again, making clear that the territory is still very much occupied. Despite the male's ambivalent attitude towards his mate at this stage it is patently obvious that he can recognise her from a distance, otherwise his über territorial instincts would kick in and he would treat her as any other trespassing Robin – in other words, drum her out!

With the weather still able to rapidly take a turn for the worse at this stage, a cold snap may be enough to briefly separate the pair, which will then revert back to their individual territories, only moving back together once the weather improves.

Breeding is at a much more advanced stage with most Tawny Owls, however, and with eggs generally laid in early March the female will now not venture far from her chosen nest site. Naturalist Dave Culley has spent hundreds of hours watching Tawny Owls nesting close to his home, and with the help of infrared cameras has been able to film behaviour rarely, if ever, seen before. Dave thinks this is the key period when the female will start to rely heavily on the male to supply her with sufficient food to acquire the necessary reserves for the rigours ahead. After dusk, from mid-February she will be constantly badgering her mate to begin the night's courtship feeding, supplemented with any food that she herself might be able to catch close to the nest site.

Also working hard to ensure their mates will be in as good a condition as possible will be the male Bewick's Swans. Amongst family parties, the obvious benefits to being dominant include access to the prime feeding areas and best roost sites, but maintaining a commanding position in the social hierarchy can also be physically demanding. At this stage of late winter, while the females continue to put on weight in preparation for both their long flight and egg laying, their partners will have often lost considerable condition as a result of putting the well-being of their families ahead of their own. It's at this stage of the year that researchers from the WWT are able to distinguish those birds that have put down the most fat reserves during the course of the winter. In slim birds, which are probably lower in the social hierarchy and last to feed, their vents (or bottoms) will look slightly concave, but the vents of well-fed birds appear much fuller, with the appearance of a 'double bulge'. In the world of Bewick's Swans, it seems a 'big behind' means you've wintered well and should carry you in good stead!

Having been happily solitary all winter, male Kingfishers will often find that the beginning of the breeding season can be a touch uncomfortable as they become accustomed to a newly arrived female. The early stages of courtship can tend to look quite similar to the threat display that Kingfishers on territory will adopt if a trespasser comes on to their patch. Announcing her presence by calling, a female's technique is to perch close to her prospective mate. This elicits an immediate response, which sees him standing upright, dropping his wings and positioning himself in such a way that she is able to see the full extent of his dagger-like bill. If the match is deemed compatible, he may then begin to make soft wooing whistles, to which she will respond with a similar call.

While male and female Kingfishers become acquainted with one another, by mid-February British Lapwings will also begin arriving on their breeding grounds. Lapwings seem to be highly faithful to the same sites year after year, and certainly in northern England have been found to nearly always return to identical or neighbouring fields. Their spring arrival is initially fairly unobtrusive, with the first parties back often preferring to feed on fields close to the ultimate breeding territories as they doubtless gauge the lie of the land. The males generally tend to arrive a week or so earlier than the females, who prefer to stay away from the melee until the territories are carved out. Preferring to visit the breeding grounds initially at dawn and dusk, the males will then tentatively start displaying, with moonlit nights often proving popular.

Those Lapwings of continental origin will begin to return to their breeding areas at this time and one of the most unexpected aspects of Lapwing migration is that, although most British-bred Lapwings will return to where they hatched as chicks, a few of these 'British' birds will undertake 'abmigration'. Abmigration is considered a northward (or north-easterly) migration, without a corresponding southward (or south-westerly) migration the previous autumn. The fact that some Lapwings seem to go in the opposite direction to what might be expected is only known because the BTO has recorded a number of Lapwings ringed as chicks in Britain which were then subsequently found breeding in countries as far away as Russia. In these cases it is suspected that British birds may well have simply migrated with birds of an eastern origin, after accidentally joining the 'wrong' flock in winter.

For those 'fair weather' British Lapwings that spent the winter in France, Spain and Portugal, their journey back to their breeding grounds will be insignificant compared to that of the Swallows currently leaving their wintering grounds in South Africa.

It is not exactly known how Swallows time their departure, but changes in daylight, the weather and food supply are all thought to be significant factors. Certainly with the northern hemisphere on the cusp of entering spring, countries such as South Africa will now be heading towards their winter, resulting in mid-February seeing the start of a mass 'hirundine' exodus northwards.

How a bird weighing little more than 20g manages to navigate its way across 9,600km of incredibly varied terrain and sea still largely mystifies the scientific community, and in many ways is as close as we may ever see to a miracle. It's now thought that migrating birds, such as Swallows, are employing a 'magnetic compass' to some extent, but this is certainly not the only cue they will use. Perhaps the position of the sun in the sky, or the pattern of polarised light on a cloudy day will also help, and they may also use landmarks, topographical features and even familiar smells when homing in on sites previously visited.

With the Swallows at last on their way, the Puffins will also finally be coming towards the end of their seven-month stint out at sea and, moult permitting, should be slowly turning their beaks towards their breeding grounds. Puffins are long-lived birds, with the oldest known bird re-caught on the Western Isles in 2012 just short of its 37th birthday. Being masters of longevity means that Puffins also have a slow adolescence, rarely reaching adult maturity until at least four or five years of age. This slow maturation means that juvenile birds will take a few seasons to develop the full characteristic adult appearance, and so young birds can easily be identified by their much smaller, less colourful bills with fewer grooves, and muted tones to their plumage. As these one- and two-year-olds will currently have virtually no chance of either attracting a mate or securing a burrow in the socially complex world of a Puffin colony, many of these 'teenagers' will simply steer clear of the colonies. Preferring to stay at sea, they will then be free to 'look after number one', biding their time until they feel ready and able to enter the cut-and-thrust world of a thriving puffinry.

Like the juvenile Puffins, Waxwings are in no hurry to leave their wintering grounds either, but for a very different reason. In common with the Bewick's Swans preparing to depart for their Arctic Russian breeding grounds, the Waxwings won't return to their northern Scandinavian summer territories while they are still snow- and ice-bound. With summer arriving much later in northern Norway, Sweden and Finland than in the UK, conditions may not even be suitable for nest building until at least early June. This means the Waxwings will still have plenty of time to track down the ever-diminishing supply of berries here before the migration instinct takes over.

Come the middle of February, a combination of data collated from satellite transmitters, geolocators and ringing recoveries all suggest that our British-breeding Cuckoos and Nightingales will both still be safely ensconced on their regular wintering territories. Even for these late departees, their primarily insectivorous diet means they should be in no immediate hurry to scramble back to a cold, wet and windy Britain. Staying where the feeding is good makes infinitely more sense, and they should be confident that when the time to leave does come, they will depart in the best possible condition.

Late February

Having spent just over the last six months as grumpy hermits, it's now time for the Kingfishers to play centre stage in the mating game. Late February and early March are great times to watch these iridescent beauties put their early un-cooperative attitudes behind them, and turn over a new leaf by entering into their new relationship with a touch more gusto. With the slightly forced introductions now behind them, the Kingfishers will begin to cement their bonds for the oncoming breeding season by conducting noisy aerial flights as they chase one another up and down their territory. These high-speed games of pursuit can sometimes last for hours, as one minute they whizz past a few inches above the water, only to return a few seconds later well above the tree tops. It can also be tricky at this stage separating aggression from romance, and it is not uncommon to sometimes see three kingfishers pursuing each other, with an interloper hoping to gatecrash the party.

Despite being so brightly coloured, many novice birdwatchers often struggle to catch sight of a Kingfisher because of the sheer speed at which they move around their territory. But the key is to listen for their shrill, tinny whistle, often given in flight, which tends to both announce their arrival and prepare you for a fly-by. Unlike many other birds, a Kingfisher's territory will be a linear one, often consisting of anywhere between 1km and 1.5km of river, stream or lake edge, and most of these courtship flights will tend to be centred around prospective nesting sites.

With the Kingfishers busily committing to one another, the male Lapwing in late February will be principally preoccupied with establishing a territory. Only when secured will he be in a position to woo a female, by demonstrating both the quality of his real estate and the calibre of his display within it. Nesting in open sites like farmland, pasture, marsh and moorland, Lapwings are on the one hand strongly territorial birds, but on the other, a species that often prefers to nest in loose association amongst its own kind. While group defence against dangerous predators like Foxes may make little real difference, an angry mob of Lapwings will be far more effective in driving away a marauding Carrion Crow from their eggs and chicks than that of a single pair nesting alone.

Lapwings have a complex series of breeding season behaviours, some of which are conducted in the air while others are confined to the ground. Territorial establishment starts with the males occupying definite positions, which will then proceed to form the core of their territory. These territories can vary in size, and Lapwing expert Michael Shrubb's studies seem to suggest that in Britain they may vary between 0.3 and 0.9 hectares. When territories are being laid down, boundary disputes are common and tend to see the rival males facing one another while puffing out their breast feathers and drooping their wings a little. Following this face-off they will then turn to walk in parallel along an imaginary line, as they conduct a staring match on the move. Early on in the season these staring bouts can also become physical as they flutter into the air while trying to drive their competitor down with a few well-placed kicks.

The most important ways for Lapwings to declare ownership, however, are with the aerial 'alternating flight' display and distinctive 'song flight', which when seen must rank as one of the great spectacles of early spring. Most common at dawn, dusk and even at night-time, the 'alternating flight' consists of the male flying in a zig-zag course, while showing his black and white plumage to its full extent, as he works the boundaries of his territory. This flight is often accompanied by a distinct thrumming noise made by the wind whistling through his outermost primaries. The male may then mix up the 'alternating flight' with his 'song flight', which comprises a rise up, followed by a steep tumbling, twisting dive. This impressive aerobatic feat is always accompanied by his characteristic '*pee-wip, wip, wip, pee-wip*' call – once heard and seen, never forgotten!

These aerial manoeuvres serve to pass on a variety of messages to any other Lapwings looking on. Firstly, they advertise not just that he is in possession of a territory, but also the size and dimensions of that territory. Additionally, because his flights are so physically and technically demanding, he is demonstrating to any neighbouring males and unmated females watching from the sidelines that he is fit, agile and a force to be reckoned with.

While continuing to dazzle with their own display flights, and just a few weeks before the female will start to produce her clutch, the pair of Peregrines will be cementing their bond with a combination of courtship feeding and ritualised ledge displays at the nest site. Co-operative hunting will now fall away as the female insists that it is the male alone who must provide her with all her dietary needs in preparation for the rigours of egg laying and beyond. Initially, he may be a touch reluctant to feed her, but as he gets his act together he will fly in with prey to where she is perched, and then transfer the meal to his beak, before passing it to his hungry mate. He will also sometimes drop the prey on the ledge after landing and retreat to let her enjoy the meal, or simply drop the prey to where she is perched as he flies past.

The couple may even undertake aerial transfers of prey, with the female leaving the building or cliff, and either catching the dropped prey in mid-air, or even more impressively, flying up below him before flipping over so that he can transfer the food from his talons to hers. With the pair now roosting together constantly, they will also frequently display to one another, particularly when close to the site chosen for egg laying. These ledge displays involve either the male or female, or even both, noisily bowing to one another in subtly different ways, which seems to vary from passive one moment to aggressive the next. After a time the pair will then graduate on to a level that can almost seem 'affectionate', as they engage in billing and gentle nibbling while producing their characteristic '*ee-chupping*' calls, reminiscent of a Herring Gull with a sore throat.

With Tawny Owls being the first of our chosen twelve to begin their clutch, the male will now be stepping up his courtship feeding duties to ensure the female acquires the necessary condition to come up with the goods. For those Tawny pairs largely dependent on small mammals, which often tend to be cyclical in population, if it's a 'bust' year then these owls may at this stage decide to abort the breeding attempt, preferring instead to hang on for another year when hopefully rodent numbers will be 'booming' once more. However, if the population levels of Bank Voles and Wood Mice look promising early in the season, then the female should be well fed, indicating that her youngsters will be well provisioned inside the nest. Perhaps most crucially of all, a good mouse and vole year will also see plenty of food on offer as the young slowly teach themselves to hunt in the summer and autumn.

Certainly by late February, any male Robin that has not yet managed to secure a mate may well be in for a barren year, unless he gets a lucky break. With Robin pairs still doing their level best to ignore each other, a small number of females may even leave their new partners at this stage, and opt for a quick divorce by taking up with a neighbouring male, or one even further away. Even though polygyny has been occasionally reported in Robins, where two females have separate territories as the male switches between both, monogamy is generally the order of the day. Experienced and established pairs will hope to raise a minimum of two broods during a good year, but with temperatures still depressed, and invertebrates – or chick food – still thin on the ground, there will still be no immediate hurry for the Robins to get down to the business of nest building just yet. After the male has gained a female his song quickly drops in intensity, and he will remain in this 'amber' state unless he is either forced to pick a fight with an intruder or if his mate is lost, so by now most of the Robin song heard should be coming from unpaired males. This behaviour doesn't just occur with Robins but also American Song Sparrows, leading the American ornithologist Margaret Nice to write: 'I often say to myself on nearing a territory where silence reigns overnight, such and such a male must be either dead or married, and upon careful searching I either find two birds or none.'

For Blue Tits aiming to time their clutches to hatch at the precise moment that caterpillars are at their most abundant, like the Robins there is currently no need to rush headlong into nest construction. Now very much an item, during the periods when the male is not preoccupied with maintaining his territory, the pair will often feed together as they visit garden bird tables or hunt through the still bare twigs for food. It tends to be around now that the sexes, when together, are most easy to differentiate. The male is usually a little larger and the blues in his plumage are a deeper colour while the yellows tend to be a touch richer. Also, in good light the male's crown, when raised, will take on a cobalt, iridescent sheen.

Still finishing their moult as they begin their epic migration, the Swallows will now be streaming out of South Africa on a broad front, hoping that if all goes according to plan, the south coast of England should loom into view in around five or six weeks' time. Swallows tend to migrate in fairly small loose flocks, meaning that individual birds are able to hawk for insects en route. Feeding on the way means that, unlike many other migrating birds, they do not need to build up large fat reserves beforehand. However, they will not be flying with the fuel tank needle hovering just above empty, as deposits may well be needed for insurance should they hit any adverse weather conditions along the way. Also, reserves will be essential when crossing inhospitable areas, such as the Sahara Desert or Mediterranean Sea. Catching insect prey as they go means that the Swallows will need to travel mainly during the day, rather than at night, with migrating birds mostly seen anywhere up to 18 metres above the ground. While on the move, they seem to prefer flying into a headwind or crosswind, which makes it easier for the birds to hunt low down without fear of stalling in mid-air.

On a good day they are believed to cover anywhere between 200 and 300km, and will try to find good roosting sites like reedbeds or dense grasslands each evening to lay up before starting out again early the following morning. Thought to travel in a north- to north-westerly direction as they leave South African airspace, late February should see our British Swallows moving through Namibia and Botswana and over the peculiar panhandle of the Namibian Caprivi Strip as they press on into Angola and Western Zambia.

With the Congo Basin still ahead of our Swallows, most British Cuckoos will still be in the Congo towards the end of February, with the majority of satellite transmitter readings centred around the mighty Congo River. The time at which they too decide to leave their wintering grounds will be quickly approaching, but unlike the 'feed on the wing' technique of the Swallows, the Cuckoos will carry out their spring migration back to Britain in a series of long-distance hops, with clear breaks to rest and feed up in between. With their departure imminent, the Cuckoos will almost certainly be trying to eat well and rest as much as possible before their first leg, which will see them travel anywhere from 500km to 2,000km in a northwesterly direction away from the Congolian swamp forests.

With the Cuckoos on the verge of leaving, most adult Puffins will be steadily working their way back to their natal colonies by late February. When airborne, the Puffin is thought to be able to cruise at around 58km per hour, and so could theoretically travel long distances quite rapidly. But with their short wings needing to beat at a frankly astonishing 400 beats a minute to keep the bird airborne, the energy demands this places upon them would require them having to regularly pitch down on the sea to replenish their reserves before carrying on. As the waters around most British puffinries won't see Puffins in any large numbers before at least the middle of March, this would suggest that arriving at their breeding grounds in good condition is distinctly more important than arriving back early.

Returning too early to their north Russian breeding grounds would simply mean that the Bewick's Swans would be unable to eat, because their breeding grounds on the maritime tundra would still be covered with snow, as well as the adjacent coastal waters probably still being frozen. Nevertheless, with the swans needing to make at least a couple of substantial refuelling stopovers before reaching the tundra, late February will see the majority of birds leave sites such as WWT Slimbridge as they conduct the first leg of their journey to the continent. The birds don't all tend to leave in one flock, but numbers will fall away as the month progresses, with a few key nights seeing large numbers leave together.

James Lees, one of the WWT wardens at Slimbridge, has been lucky enough to see a few large departures over the years, and says that a few days before the swans depart they become decidedly nervous at feeding time. They will also spend substantial parts of the day sleeping, no doubt to conserve energy for the flight to come. The majority of the swans will only leave when the conditions are right - favouring dark, clear nights with plenty of stars to aid in navigation and a westerly wind to help push them in the right direction. When the precise moment for departure comes, it is usually preceded by the birds becoming incredibly vocal, with a huge amount of head bobbing. Only when the noise level and degree of synchronised head bobbing reaches fever pitch will the flock rise from the roost, before spiralling into the air to quite a height, in order to gain their bearings, as they head off for the continent.

In a good Waxwing year, the number of birds in Britain by now may well have declined from the peak seen at the turn of the year. However, late February may well still see these gentle marauders dispersed far and wide as they track down the few remaining supplies not already mopped up by the resident Blackbirds, Song Thrushes, Mistle Thrushes or the combined total of around 1.5 million Redwings and Fieldfares also visiting Britain for the winter. Even in a year when relatively few Waxwings cross the North Sea, the east coast at this time will almost certainly see a smattering of flocks visiting any garden with berries going to spare. Always approachable, probably due to the lack of human interference on their breeding grounds, the Waxwings will still be in no hurry to leave.

Out of our four summer visitors, the only species that has seemingly not even entertained thoughts of leaving come the end of this month is the Nightingale. Recent data collected from the BTO's research work suggests that the birds will not leave until mid-March. But by feeding well in West Africa, once the Nightingales finally do decide to leave, the first leg will see them power to southern Europe with all possible haste.

March

Even if the weather conditions seem to say otherwise, higher daily temperatures and a few precious minutes of daylight clawed back with each passing day can only mean spring has most definitely sprung by March. As a yellow army of Daffodils, Primroses and Lesser Celandines begin to appear and overwintering butterflies and queen bumblebees tentatively emerge, many of our winter visitors will take the improving conditions here as their cue to desert our shores for long, arduous journeys to far-flung locations. The noisy and extensive dawn chorus here in March, however, is ample demonstration that this month has already become a hectic time for our resident British birds. For any individuals that successfully negotiated the winter, their reward will have already come as they find themselves either on or close to their breeding grounds, putting them in pole position to race ahead in the mating game. However with most summer migrants well on their way, the resident birds will be only too aware that they won't have our gardens, woodlands, farmland, scrub and wetlands all to themselves for long! Despite many of our summer visitors still being thousands of kilometres from Britain, the traditional advance party of Wheatears, Sand Martins and Chiffchaffs will have already begun to arrive at many southern migration watchpoints during this month. In fact, this initial trickle will soon turn into a ranging torrent as millions of birds head towards Britain to take advantage of our green and pleasant lands.

Early March

The first of our chosen twelve to kick-start the egg-laying season will be the Tawny Owls. Being a cavity-nesting bird, the female Tawny doesn't need to produce a camouflaged clutch, and so her rather round eggs are white and unmarked. Before laying each egg the female tends to become distracted, and using her bill to grasp on to the walls of the nesting chamber, the egg will often be delivered while standing up. Tawny Owl expert Dave Culley has spent many sleepless nights waiting for his wild Tawnies to produce their clutch, and reckons the female will lay an egg every 56 to 64 hours, with sometimes a delay of up to four days before the final egg is produced. Clutch sizes can vary, but two or three is considered the norm, with some pairs capable of producing four or even very occasionally five. The time-consuming matter of incubating the clutch is carried out by the female alone, and from Dave's studies, he reckons she will only properly begin sitting with the penultimate egg. In preparation for the 'big sit' she will also pluck out her breast feathers to reveal the brood patch, a bare area of skin well supplied with blood vessels at the surface, making it possible to easily transfer heat to the eggs. If she has to leave the clutch for any reason before proper incubation commences then she will cover the eggs over. From the point incubation begins she will suddenly become utterly reliant on the male provisioning her with food during this period of solitary confinement.

Only marginally behind our Tawny Owls in the laying stakes will be the Peregrines. At this stage, the male will be taking full advantage of his conjugal 'rights' by copulating with his mate as frequently as possible. Most female Peregrines are substantially larger than their partners, so the male needs to be careful to telegraph when he'd like to mount her, to ensure his actions are not misconstrued as aggressive. Stating his intention involves a ritualised hitching of his wings, making it look just like he's shrugging his shoulders and can sometimes be accompanied by the male prancing around as if on tiptoe. If the female is agreeable to his overtures, she will then lean forward, and swing her tail to one side, inviting him to jump on. Resting his tarsi on her back and with his talons curled into a ball to make sure he doesn't harm her, he will then press his tail under hers. Constantly flapping his wings to maintain balance and ensure he doesn't slip off, the entire mating deed will often take less than ten seconds. This briefest of liaisons also tends to be conducted amid a cacophony of noise from both the birds.

Despite moments when they're still a touch uneasy with one another, many pairs of Kingfishers should by now be fairly established and settled for the oncoming breeding season. Once a partner has been secured, the first priority for the newly engaged couple will involve choosing the appropriate bank in which to excavate a nest. Kingfisher nests are always situated in an enlarged chamber at the end of a narrow tunnel, and unless the pair decides to refurbish an existing 'second-hand' nest, they will dig a fresh one. An ideal nest site should be easy to excavate, yet safe from both predators and flooding. Most sites chosen tend to be in vertical, steep or overhanging banks on the edge of the watercourse, and are often sited both at least a metre above the water and 50cm or less below the top of the bank. Crucially, the location must also be within the birds' territory. Soft or sandy soils with few roots and stones will make the digging more straightforward.

Excavating the tunnel and chamber is generally a shared task, but the males will do the lion's share of the digging. The hole is initiated by the male flying at the bank with his bill extended, while his mate watches from close by. The moment a foothold has been created, the Kingfisher is able to enact an impression of a woodpecker, as he wields his bill like a pickaxe while hanging on to the bank. During the whole tunnelling process, the pair often work in shifts, with the non-digging bird on the lookout for both intruding Kingfishers and possible predators. Most of the digging tends to be carried out in the morning, and as the tunnel gets longer, the process seems to become easier. The digging bird will use its feet to push the soil behind it, and as the tunnel measures little more than 5 or 6cm across, it then has to reverse out, using its tail feathers as a mini-bulldozer to clear out the spoil. The tunnel leading to the nest chamber is usually around 45 to 90cm long and dug either on a slight incline or in horizontal fashion, and only when work starts on the actual nest chamber will the birds finally be able to turn around inside and emerge from the tunnel bill first. Depending on how hard the soil is, and whether any obstructions are encountered along the way, the whole process may take between a week and 12 days.

In contrast to the engineering feat that is a Kingfisher's nest, Lapwing nests consist of little more than a scrape in the ground. This arguably gives the males far more time to perfect their wonderful flight displays to both impress any as yet unpaired females, and to warn any neighbouring males against trespassing. As many Lapwings choose to nest in 'loose colonies', polygyny is considered quite common, with one study in Teesdale, County Durham revealing that between 20 and 44% of males had been successful in securing two or even three mates. Those males with two or more females often tend to be more experienced, and so have larger and better quality territories more capable of rearing multiple broods simultaneously. It's also thought that many females may ignore the younger males, preferring instead to playing second fiddle to an already mated male. This mating system inevitably means there will be a number of males, holding possibly peripheral and inferior territories, which may miss out on a mate entirely. For these birds, the breeding season may well then be downgraded to a watching and learning brief, with the hope that any experience banked will count in their favour in future years.

At this stage, neither Robins nor Blue Tits will have advanced their breeding season any further than establishing a territory and attracting a mate. As invertebrate prey is still low in abundance at this time of year, and with incubation only taking around a couple of weeks, both species are in no hurry to press forward come early March. The male Robin, having secured his mate, is still doing his level best to ignore her, and when not feeding, the vast majority of his time will still be taken up by singing and fighting. Despite their territorial nature, Robins will frequently enter neighbouring territories, particularly in search of food. These trespassing Robins will rarely sing or display and try to remain as inconspicuous as possible, being only too aware that if spotted they will need to beat a hasty retreat. With the incumbent male winning the majority of his battles, the only Robin left entirely in peace within the territory will be his mate, whom he should by now be able to recognise from quite a distance. While not as strongly territorial as the Robin, the male Blue Tit will still be clearly demarcating his territory through singing and aggressive behaviour towards any intruding males. Their pair bond, which formed in winter, will also continue to be cemented with nest building now just a few weeks away.

By early March most Bewick's Swans will have left Britain for the near continent on the first leg of their epic spring migration. Flying at speeds of between 46 and 70km per hour, depending on the prevailing weather conditions, they are not thought to fly at particularly high altitudes when migrating. It is also assumed that individual birds or family parties may well use the same staging sites each year where possible, with several locations being incredibly important for refuelling along the way. It has been calculated from energetic studies that large birds like Bewick's Swans may only be able to fly an estimated 1,500km before needing to refuel for the next stage of their journey. Because the breeding sites are 3,000 to 3,500km away, the swans are thought to make at least a couple of stops during spring migration, and possibly even more if they encounter severe weather conditions along the way, or the adults are still accompanied by their offspring.

Satellite tracking of Bewick's Swans has recently been initiated by scientists at the WWT. Eighteen swans were caught between 2012 and 2014 at Slimbridge and the Ouse Washes to have transmitters attached in the form of collars around their necks. Already the data the swans are sending back to the team is revelatory, proving that upon leaving

Britain, the majority of tagged birds immediately relocate to Denmark. This research has also revealed that a proportion of these tagged birds (and therefore presumably a proportion of the population), will then continue either south and inland, into Germany, or further east into Poland. It's thought the birds staying in Denmark may well be feeding in agricultural fields, such as maize stubble, in contrast to the birds visiting Germany that are feeding in wetlands, swans in Poland which have been seen feeding around fish ponds.

Unfortunately there isn't the same level of detailed knowledge about the movement of Waxwings, but in a good Waxwing winter, both the number of birds and flocks will by now have begun to reduce as birds commence crossing back over the North Sea en route to their breeding grounds in northern Scandinavia and the Russian Federation. With spring arriving late in the boreal forests, it is reasonable to suggest that the Waxwings may well cover the 2,000km return journey at a leisurely pace, feeding along the way. Even in winters with relatively few Waxwings visiting Britain, it is still highly likely that early March will see a scattering of Waxwing records, principally along our eastern seaboard.

Of our four summer visitors, the species undoubtedly closest to its breeding grounds in early March will be the Puffin. Arrival dates tend to vary widely from year to year at the same breeding location and also between different colonies, but if each of the major British breeding sites is represented as the hub of a bicycle wheel, then the freshly-moulted adult Puffins will currently be arrowing in, like the wheel's spokes, from all directions. It is highly likely that the Puffins will also be pausing for regular feeding bouts along the way, and having spent the winter probably feeding alone, will undoubtedly have begun to meet other Puffins returning to the same colony. These encounters out at sea after a long winter of solitude should to some extent prepare them for what will be a very intense social experience ahead.

The Nightingale has a shorter distance to cover between its wintering grounds and breeding site than either the Cuckoo or Swallow, so it's perhaps no surprise that they will be the last birds to leave their final wintering destination. So, come early March they will still not have embarked on the first leg of their 5,000km journey, which will ultimately see them arriving in south-east England in mid-April. Capable of covering a huge distance in their first long hop, British Nightingales will currently be busily laying down fat stores, only too aware that running out of reserves while circumventing the Sahara Desert could be a death sentence for a bird weighing little more than 20g.

Finally deciding that they can't delay any longer, early March will see our British-breeding Cuckoos at last leaving the Congo region. Data from the BTO's work tracking Cuckoos by satellite seems to indicate that come early March most tagged birds will start heading north. What factors make them suddenly up sticks and leave behind the sanctuary of their wintering quarters is anyone's guess, as day length and temperature don't vary much throughout the year close to the Equator. The cues, however, may include changes in rainfall patterns, which will have a knock-on effect on the prey they've been consuming. Leaving the massive Congo forests behind, the general trend sees the Cuckoos pitching up in a wide belt of land consisting of mostly woodland savanna between the Central African Republic and Cameroon. After little more than a few days' rest, the majority of tagged Cuckoos will then quickly move west through Nigeria, Togo and Benin as they head towards Ghana.

In contrast to the Cuckoo's long hops with rests for feeding up in between, the Swallows will be continuing their migratory routine of travelling anywhere between 200 and 300km during daylight hours, before then finding a safe spot to roost each night. Working their way steadily north, early March may well see the Swallows approaching the Congo forests that the Cuckoos have only just left. It's thought that the Swallows may well take a more westerly route than where the Cuckoos overwintered, as continuous forest cover is probably a more difficult habitat for them to feed over than the lowland wetlands and farmland found closer to the coast. As mid-March arrives many British-bound Swallows will be steaming across the Equator and arriving in the unattractively nicknamed 'armpit of Africa' as they cut the corner of the Gulf of Guinea into Cameroon and Nigeria.

One site where over a million Swallows bed down for the night on their journeys both north in the spring and south in the autumn is close to the village of Ebok-Boje, of the Cross River forest region in south-eastern Nigeria. Roosting on hill slopes covered in three-metre-high elephant grass, huge numbers descend at dusk, only to leave at sunrise the following morning. This roost was unknown to the outside world until 1987, and quickly made the headlines as local communities were thought to be regularly trapping and eating as many as 200,000 birds each year. Since this discovery, considerable work has been put in by conservation agencies keen to initially reduce and ultimately prevent the industrial slaughter of the Swallows roosting here. Any Swallows escaping the liming sticks of Ebok-Boje will continue their journey north only too aware that an even bigger potential death trap, the Sahara Desert, will soon be looming very large on their horizon.

Mid-March

After spectacular aerial displays, a whole array of ritualised courtship and a fair amount of copulation, mid-March will see British Peregrines begin to start laying. The famous naturalist Derek Ratcliffe, in his peerless monograph *The Peregrine Falcon* in 1993, wrote that the earliest date he could find in the literature relating to when Peregrines laid their first egg was 23 March. However, the BTO has suggested that Peregrines are now laying on average nine days earlier than when Ratcliffe carried out his research. Ratcliffe also scarcely mentioned urban-nesting Peregrines in his book, as this is a phenomenon that has only been widely observed in Britain over the last 15 years. It now seems that the 'heat island effect' commonly found in towns and cities is the main reason why populations of Peregrines in urban locations may well be breeding earlier than on the cold Cumbrian fells where Ratcliffe watched his falcons. Suffice to say, there will of course be variation between different pairs and different years, with the temperature in spring thought to be the key factor in dictating when Peregrines begin nesting.

In contrast to many birds, the Peregrine's nest is always a relatively simple affair, consisting of little more than a shallow depression with a few pellets and feathers drawn together. With the precise nesting location key to the success of the birds, 'Peregrine boxes' on churches and other high buildings have proved spectacularly successful in cities such as Bath, Norwich and Chichester. Here the box often consists of little more than an open-sided rectangle, with a roof to offer shelter from the worst of the rain and pea gravel as a substrate on which to nest, with clever placement ensuring that a sudden gust of wind won't blow the precious clutch either off a ledge or into a gutter.

The sign that a female may well be just about to begin laying will see her fairly restless early in the morning, prior to the egg being produced. This egg will represent the first in a final clutch of three or four, with as many as five reported by Peregrine ringer Ed Drewitt in Bristol's Avon Gorge in 2008, 2010 and 2011. The prevailing colour of the eggs is red-brown, with varying degrees of freckling, mottling and blotching, and an unashamedly biased Derek Ratcliffe called them 'the most handsome laid by any species of bird'. The eggs do, however, develop a 'shop-soiled' look during incubation as they become dirtier and splattered with droppings. A clutch of three or four may take a week for the female to lay, with incubation not thought to start until the final egg has been laid. Up to this moment, any development of the embryos inside the eggs will remain in a state of suspended animation until the warmth from the parents kick-starts the development process.

Already a week into incubation, the female Tawny Owl will be approximately a quarter of the way through her vigil, with the only break in monotony occurring when she has to defecate, or her mate brings food to the nest site. Carefully turning the eggs regularly to ensure they are evenly warmed, she is every inch the devoted mother. At this stage, many female Tawny Owls will actively defend their clutch from what they consider to be potential predators, with a number of anecdotal reports of certain individual birds attacking mammalian intruders such as dogs and Foxes encountered too close to the nest sites. In reality, only some will attack, as many sitting females may prefer to sit tight, while others either slip away or even become the model of docility. It must of course have been a female at the aggressive end of the spectrum that caused the famous wildlife photographer Eric Hosking to lose an eye when he was attacked entering a hide close to a Tawny Owl nest in 1937.

The only other animal likely to incur the wrath of an established pair of Kingfishers during the digging stage will be an intruding member of the same species. During this key period any trespasser 'fishing around' will be given short shrift, particularly when caught so close to such a valuable commodity as a good nesting bank. In addition to digging, courtship feeding will by now have become a regular feature in the birds' daily routine. With the Kingfishers still buzzing around their territory, the male will often follow up these high-speed chases with a fishy gift. Offering a recently caught fish with its head pointing out, as opposed to head-first when feeding himself, he will present his gift amid much noise and ceremony. Quivering her wings as she accepts the fish, which in many ways mimics the behaviour of hungry baby birds, she will be helping to ensure she is in the best possible body condition with the time for egg formation rapidly approaching.

Certainly in southern Britain, mid-March should see the Robin pairs putting their initial estrangement behind them and pulling together as the mating process moves to centre stage. Nest building is an activity undertaken purely by the female, and coincides with a rise in her aggression, as she backs up the male in repelling intruders of either sex attempting to disrupt their party. With material gleaned from around the territory, the nest is built mostly of moss on a foundation of leaves, which is then often lined with hair. The location slated for construction is usually on or near the ground and well concealed in any hollow, nook or cranny, such as in amongst climbing plants, tree roots, or piles of logs. Open-fronted nest boxes can also be attractive to Robins provided they're well concealed. Robins are also well known for nesting in all kinds of unlikely locations, and renowned Robin expert David Lack collected reports of pairs having nested in letterboxes, old boots, coat pockets, under car bonnets and even inside a human skull!

The female is very discreet while constructing the nest and will only build for around four hours each day. At this stage and during laying, the birds are very sensitive to any disturbance and will readily desert the nest if they think it's been discovered. So unless the birds are particularly accustomed to people, it's best to stay away from the immediate vicinity of a Robin's nest until the female has begun incubation. During the nest-building stage the pair may also fleetingly display to one another as a prelude to sex. During this courtship the female will suddenly remain still before lowering her head and causing the body to hunch. The male will then occasionally sway before briefly mounting her. Sex amongst Robins seems to be nothing more than perfunctory and is only performed a few times a day during nest building and egg laying.

With most Robins aiming for two broods, and possibly even a third in good years, it is in their interests to begin the breeding season as soon as the weather permits. However, as Blue Tits will aim for just one large brood in all but the most extraordinary circumstances, timing is everything. With invertebrate prey still relatively scarce for a few weeks yet, the Blue Tits certainly in southern Britain may well be holding off nest building until the end of this month. With spring arriving first in the south-west of Britain, before then proceeding at walking pace across the country in a north-easterly direction, it's no surprise that northern Blue Tits may well delay their nesting attempts by as much as a month compared to their southern cousins. As with the female Robins, it's vital that the female Blue Tits remain fit and strong in preparation for the prodigious toll egg laying will exact on their bodies. Begging regularly from her mate at this stage, the female will indicate her desire to be fed by fluffing out her feathers, lowering her head and fluttering her outstretched wings, behaviour that will carry right through until when the chicks hatch.

As the female Blue Tits eat almost enough for two, the last unpaired Lapwing females should now be making their final selection of mate and territory. The females won't just be admiring the males' dazzling displays, but will also be running a keen eye over the quality of their real estate. The best territories have features that help camouflage their nest, such as broken or cryptic backgrounds, and are placed well away from trees and field boundaries so as to deter perching predators. Ideal locations will also be sited close to good feeding or chick-rearing areas, such as tilled land or wet areas, and additionally contain some short vegetation or bare ground on which to nest. Once the females have moved into the territory of choice the next step will be to cement their pair bond with courtship displays.

For the Puffins returning to their nesting colonies, renewing acquaintances will be a high priority before they even think of putting a foot on dry land. The first records of Puffins arriving back in spring are always of small numbers of birds on the water close to the colonies. Different colonies tend to receive the Puffins back at slightly different times, but the pattern of events at sites such as Skomer Island, the Farne Islands and St Kilda is always similar. The first birds arriving back will do little apart from float around in small groups facing into the wind and waves, but as numbers steadily build up, and established pairs meet up again after a winter in all probability spent apart, they will renew their wedding vows and pick up from where they left off last August. The most frequent early demonstration of an established pairing is 'billing', where the Puffins will rapidly and noisily knock their bills together, with the male invariably being the one to instigate this wooing. It is not until substantial numbers of birds build up on the water that an element of 'safety in numbers' will find Puffins comfortable enough to leave the sanctuary of the sea.

Until very recently, the details of not just where Nightingales overwintered in Africa, but also when and from where they left to return to northern Europe, was still largely unknown. However, thanks to the use of tiny geolocators by the BTO, these mysteries are slowly being revealed. The first Nightingale to successfully bring a geolocator back to Britain with meaningful data, imaginatively entitled OAD, revealed the path down to West Africa, but the equipment frustratingly failed around February. This meant that the return journey was still pretty much guesswork until other Nightingales successfully returned to England with fully functioning equipment in subsequent years. Five Nightingales with geolocators attached in Orlestone Forest in Kent in 2012 successfully made it down to Africa and the data from the recovered tags revealed they all began their return migration the following spring between 11 March and 19 March.

The most remarkable of these five birds was 'Nightingale 098', which after leaving Sierra Leone took just three days to travel, in what must have been non-stop flight, all the way to southern Portugal, an astonishing distance of around 3,300km. Flying at a scarcely believable 45km per hour, and so almost certainly aided by a tailwind, the bird is thought to have taken the coastal route, passing through Senegambia, Mauritania, Western Sahara and Morocco, before crossing the Mediterranean into southern Portugal. Migrating along this route means that Nightingales are believed to circumvent the Sahara Desert, arriving in southern Europe just before their fat reserves run critically low – all in all, a pretty impressive feat of both physical endurance and navigational ability!

While our Nightingales are busy leapfrogging their way to southern Europe, mid-March sees most British-breeding Cuckoos resting and feeding up at a diverse range of locations centred around Ghana. A number of the satellite-tagged birds, including 'Cuckoo Chris' have at some point dropped in to Digya National Park, situated in the east of the country, but on the west bank of Lake Volta, the largest reservoir in the world. Digya is the second largest national park in Ghana, and situated on a lowland peninsula that mostly consists of transitional habitat between forest and savanna. So having spent the winter possibly feeding alongside gorillas, the Cuckoos will now be fattening up close to elephants as they prepare to negotiate North Africa and the mighty Sahara.

Further north our intrepid Swallows will have just crossed the semi-arid habitat of the Sahel, a region that straddles the

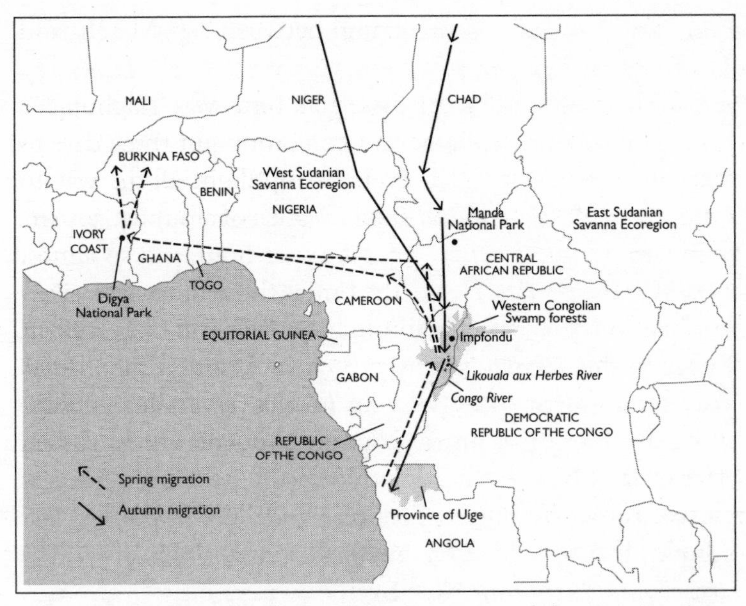

Both the spring and autumn migratory routes and the overwintering destinations of British-breeding Cuckoos in Africa.

entire continent in a broad band between Senegal in the east and Ethiopia in the west. So by mid-March they should be girding their loins in preparation for taking on the largest hot desert in the world. Extending over nine million square kilometres, and covering a quarter of the entire African continent, the huge landforms of the Sahara Desert are shaped by both the wind and rainfall from a more fertile era. Consisting of sand dunes, sand seas, stone plateaus, dry valleys, dry lakes and salt flats, the Sahara Desert has to be one of the most inhospitable places on Earth. It is also believed to he too much of a detour for the Swallows to follow the return route of the more westerly-wintering Nightingales, meaning they will have no choice other than to engage the desert head on if they're to reach the insect-rich summer of northern Europe. Having fed the whole way since leaving South Africa a month before, it's highly likely that the lack of vegetation, relentless sun and strong winds will mean that invertebrates over the desert are few and far between. So as the Swallows pass through Mali and Niger and on to the huge desert country of Algeria, they will just have to trust that their limited fat reserves will see them through beyond the sand to the rich coastal crescent of North Africa.

While the Sahara Desert and the Mediterranean Sea are reached much earlier during the southbound migration in autumn, the return spring journey is a different story. Travelling north, the two immovable obstacles of sand and sea will come much later in the journey, and only after the birds have already flown several thousand kilometres across Africa. Doubtless the passage across this hugely inhospitable terrain must be physically very stressful, with innumerable Swallows perishing in the sand before the stronger and more experienced birds finally straggle into the coastal feeding areas of Algeria, Tunisia and Morocco.

There are few recorded sightings of Swallows in the Sahara itself and the 1,500km distance across the desert expanse will surely be the most gruelling five or six days of their year. Possibly flying for up to 14 or 16 hours a day, the Swallows must presumably roost wherever they are able, before the breeding imperative drives them on at dawn the following day.

Of the total of 35 Waxwings caught and ringed while overwintering in Britain only to then be subsequently recovered abroad, 27 were found across Denmark, Norway, Sweden and Finland. The fact that these birds were found in a diverse array of locations across Norway, Sweden and Finland is perhaps no surprise, as these will almost certainly be records of birds either close to, or at their breeding grounds, but the six from Denmark is in many ways more revealing. Waxwing as a species doesn't breed in Denmark, so this would suggest that any British-ringed birds observed must be en route, thereby revealing the main route that returning birds will take on their way to northern Scandinavia and the Russia Federation. With many Waxwings already having left it's perhaps surprising, even in a poor year, how a number will still be happily to feeding in Britain late into April. These late leavers must then presumably make a much quicker return to their breeding grounds than those that left earlier in the year.

From now on, Britain should be a Bewick's free zone until the first birds begin returning in the autumn. Having reached Denmark, Germany, Poland and southern Sweden in one hop it seems that many of our Bewick's are in no hurry to reach the Russian maritime tundra, which will certainly still be frozen over at this time of year. With seemingly plenty of food still available in northern Europe, satellite tracking data from the WWT indicate that some of the Bewick's Swans will not have moved far from their first arrival point. Other birds, however, will have begun to slowly spread out along the Baltic coast, passing through Poland, Lithuania and Latvia on their way to internationally important stopover locations in Estonia.

Late March

Finally emerging from the Sahara, only to then hit the Atlas Mountains, extending some 2,500km in a south-westerly to north-easterly direction through Morocco, Algeria and Tunisia, conditions for the Swallows should suddenly begin improving as they catch sight of the Mediterranean Sea way off in the distance. As the birds drop down towards the coast, the Aleppo Pines and Evergreen Oaks of the mountain forests will soon give way to Mediterranean-type scrub, called maquis, and a welcome change of climate. For those Swallows that haven't either perished in dust storms, or fallen from the sky due to exhaustion or a lack of water, the cooler temperatures and higher rainfall encountered all along Africa's north coast will not have come a moment too soon.

Exhausted from their efforts, and with many Swallows having lost a third of their body weight, they will need to feed up quickly before then crossing the Mediterranean Sea and arriving in Europe along a broad front, anywhere from the eastern coast of Spain to western Italian shores. Finally back over dry land, for those birds that took the more westerly

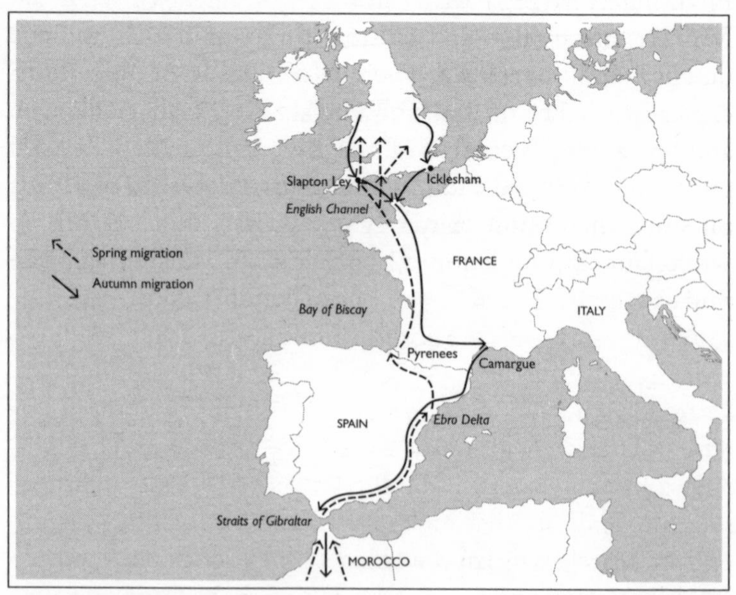

The believed spring and autumn migratory routes of Swallows between their breeding grounds in Britain and north Africa.

route, sites such as the Ebro Delta, situated between Valencia and Barcelona, will represent an important refuelling station. Quickly moving on towards the Pyrenees, these Swallows will then begin streaming across the autonomous community of Aragon in the north-east of Spain. With Aragon's northern boundary forming the border with France, and positioned in the middle of the mountain chain, the Aragonese say that 'the Swallow is the bird that melts the snow', as the birds work their way east to west, through mountain valleys flushed with melt water. By the time they reach the Bay of Biscay and the French coast, the Swallows should be spurred on in the knowledge that they're on the home straight.

Able to help themselves to 'in-flight food' the Swallows are in the fortunate position of being able to press on, but any Nightingale having passed through West Africa and over the Mediterranean in one non-stop flight will need to feed quickly upon reaching land, before its fat reserves run critically low. The dry, uncultivated land in southern Portugal and Spain, which consists of a mosaic of scrub and wooded groves of Cork and Holm Oak, is thought to be the main pit stop where most of our Nightingales will then spend a while fattening up before ultimately leaving for England in mid-April.

The only one of our chosen migrants still overwintering in Africa by this stage, will be the Cuckoo. Satellite telemetry seems to indicate that most British-tagged Cuckoos by the end of March will have worked their way steadily west to reach countries such as the Ivory Coast. Traditionally very heavily forested, particularly in the lowland Guinean forests along the coast, there has been much recent concern about the rapid rate of deforestation in this old French colony, sometimes cited as the highest in the world. Certainly 'Cuckoo Chris' has spent two of his three pre-Saharan stopover points in March feeding up in the eastern part of the country. Fortunately the area Chris and the other Cuckoos seem to favour is in and around the forest-savanna mosaic, a belt extending across the centre of the country, and to the north of the most heavily deforested areas. Once the Cuckoos do leave this habitat in early April, they will in all likelihood not touch down again until they arrive in southern Europe. In the meantime, the only question vexing the Cuckoos as they prepare to traverse the huge arid zones further north is which route to plump for when they cross the Sahara Desert.

By late March, most inshore waters around the major colonies will have begun to turn into a Puffin-thronged melee. Emboldened by their numbers, it seems surprising that the Puffins don't immediately pour on to the breeding colony, but instead appear reluctant to leave the sea. During the first few days back around the colony, the Puffins will use this time to join up in rafts with those birds that have become close neighbours during previous breeding seasons. These 'clan' gatherings at sea often tend to become aligned adjacent to the sections of the colony with which they're most familiar, and small flocks will then quickly coalesce into much larger rafts of possibly several thousand Puffins as more arrive back. Suddenly amid much excitement, a large number of Puffins will then begin taking off together before heading for land. Flying towards the puffinry, the Puffins will then circle over the area where they intend to breed, before looping back over the sea. This 'circuit' is flown a number of times, with the number of birds rapidly increasing all the time until the flock becomes a ring of fast-moving

Puffins called a 'wheel'. The different 'sub-colonies' within the puffinry will often have 'wheels' in different locations, enabling the birds to check out familiar terrain while in a large, speedy flock.

This important event in the life of the Puffin is much more than just a convenient way to travel from sea to land, as it also provides an ingenious method of checking for predators close to where they intend to nest. It's also far more difficult for any predators, such as Great Black-backed Gulls, Peregrines or Great Skuas, to pick out just a single bird to target for attack from amongst a rapidly swirling flock. Eventually a few brave Puffins will peel away from the wheel before touching down on terra firma for the first time in over seven months, which in turn gives others confidence to join them on dry land. It's only when this 'touchdown' has occured that the Puffins' breeding season could be said to have well and truly started.

The two birds with a considerable distance still to cover before they even reach their breeding grounds will be our two winter visitors, Bewick's Swan and Waxwing. Those Waxwings that have already left Britain will in all likelihood have crossed the North Sea and be currently spread out at still fairly low latitudes in a broad front across Denmark, Norway and Sweden. As they head north-east towards the northern forests surrounding the Arctic Circle, they will certainly be eating on the way. Due to the breeding season at high latitudes often being short, many Waxwings will be keen to hit the ground running the moment the weather ameliorates. This ability to quickly capitalise on conditions should give them sufficient time to successfully raise chicks before the days once again shorten and temperatures plummet.

Thanks to intense field work by researchers from the WWT and recent advances in satellite telemetry, the migration picture is currently much clearer for Bewick's Swans than for

Waxwings. As the month draws to an end, most British Bewick's Swans will suddenly leave Denmark, Sweden, Poland and Germany to head towards their key spring staging posts – the estuaries and wetlands off the coasts of Estonia and Latvia, most of which should by now be ice-free. Matsalu Bay, and Pärnu just further south, have long been recognised for their importance for migrating Bewick's Swans, with Matsalu being re-designated as a Ramsar site by the Estonian Government in 1994. These sites were well watched between 1992 and 1994, with Matsalu accounting for approximately 14,500 Bewick's Swans and a further 7,000 dotted around Pärnu Bay and the Pärnu River's watershed. When combined, this figure will have formed a very high proportion of all the Bewick's Swans from the European flyway.

Having been sitting since early March, the embryos inside the eggs should by now be developing quickly providing the female Tawny Owl has been regularly fed by her mate. With her clutch well defended, the only issues that could prejudice the welfare of the clutch at this time is if the Owls have chosen to breed in a year when prey is scarce, or if hunting activity is reduced by prolonged poor weather. In these situations the female may have to break away from incubating to go hunting herself, which could put the eggs at risk of chilling. Dave Culley has found that his resident male Tawny can be incredibly resourceful at this time of year, and with small mammals in lower abundance in early spring, frogs frequently start to form a substantial component of the items brought in to the sitting female.

By the end of March most urban Peregrines should have been sitting on a full clutch for at least a couple of weeks.

Unlike the Tawny Owl, both male and female Peregrine will incubate the eggs, although the female will invariably undertake the greater share and additionally cover the night shifts. Producing three or four eggs is an incredible investment in time and energy for the female, and so if the weather does take a turn for the worse, the incubating birds will often sit tight to ensure their precious clutch doesn't chill. A supreme example of adult Peregrines' devotion to their eggs occurred with an established pair nesting on Nottingham Trent University's Newton Building in late March 2013. An unseasonably late snow flurry virtually covered both the open nest box and the female sitting on a clutch of three eggs. Despite being virtually buried until the snow melted, she didn't abdicate her responsibility and all three chicks successfully hatched at the end of April, a testament to her dedication.

With a territory secured and one or even more females interested, the Lapwing's breeding season will now be quickly gathering pace in the more southerly breeding sites. Having taken to the air to display the calibre of his site and his qualities as a potential partner, the male will now capitalise on his good early work to further court the female on the ground. The main ground display between the pair centres on 'nest scraping'. Here, the female squats down, often in the location where she would like the nest scrape to be positioned. The male will then approach her while alternately flashing his orange undertail coverts and white rump. The female responds to this overt display by moving towards him, and bowing steeply as she too displays her bright undertail and rump colours. Moving back into the scrape the female then starts raking the ground while the male flicks any material close by along her flanks, and if suitably impressed she will then invite him to mate with her. Mating directly from flight, the male will need to use his wings to maintain his balance while copulation takes place.

Favoured locations for nest scrapes tend to be poor, stony areas with limited plant growth, which offer both good cryptic camouflage and yet excellent all-round visibility. With the precise spot chosen, both sexes will then enlarge the scrape and proceed to line it with whatever is close to hand. As she becomes broody, the female will then constantly add extra material to the nest as a precursor to laying the first egg in her clutch of four.

Once the nest excavation is complete, the Kingfishers should be in a position to mate. However, the female might not have always read the script and even at this stage can still prove unreceptive to his advances. In contrast to the Lapwings, where courtship seems to be key before mating, the male Kingfisher's tactic involves bribery. Offering the gift of a fish may well help break the resistance of any females still reluctant to commit. Suitably assuaged, the submitting female will then lie prostrate, quivering her wings and suddenly changing her tune to one of encouragement. Jumping on and flapping for balance, the male will often hold his mate's crown feathers in the tip of his bill to ensure compliance as they touch cloacas. The act itself often looks quite rough but is usually over in less than ten seconds. Unlike the Lapwings, the Kingfishers don't collect any material to actively line their nest, but a smelly layer of fish bones and scales will frequently quickly build up to make a cup of sorts as the breeding season develops.

Even though it's doubtful the male Robin will ever win any awards for 'most devoted husband', his generosity does extend to helping feed his mate as her nest building nears completion. Courtship feeding is often initiated by the female uttering a sharp call as the male approaches with food. Lowering her wings and quivering with excitement, her tune then changes to a rapidly repeating note, mimicking the call of a hungry baby bird. By constantly begging from the male in this way she will receive the vital calories necessary for the incredibly demanding task of producing a clutch amounting to two-thirds of her body weight in the space of just a few days.

As spring arrival in Britain varies according to location, and with Blue Tits keen to rear their young at a time of maximum food abundance, the dates when a pair starts nest building can vary enormously according to whether they are Blue Tits from Devon or Dundee. Generally in southern England, Blue Tits will begin construction duties in late March, with the same chore often pushed back by as much as a month in northern Scotland. In addition to this, sudden weather changes such as prolonged wet periods or a warm spell can even produce variation from year to year at the same location. The job of building the nest is always carried out by the female, and so the period when March fades into April should see many female Blue Tits gathering the necessary nest construction material as they start the ball rolling on their one and only breeding attempt of the year.

April

With around six hours more daylight and average daily temperatures at least 5°C higher than the dark, cold days of January, April is the month when the natural world breaks free from the constraints of winter, and attempts to make up for lost time by forging ahead with the business of passing on its genes. As frogs and toads squabble for mating partners, hedgehogs, bats and reptiles will also have fully emerged from hibernation by now. This frenzy of activity also coincides with the buds of familiar trees such oak, Beech and Hawthorn bursting as Bluebells, Wood Anemone and Wild Garlic carpet the woodland floor below. This month will also see the majority of migrant birds pouring back into Britain, which will instantly transform the dawn chorus into a sound sensation as, depending on the habitat, birds from Blackcaps and Bitterns to Woodlarks and Willow Warblers add their voices to this glorious natural symphony. Stealing a march on these fair-weather immigrants, many of our resident birds should by now either have started to lay, or already be incubating clutches as they attempt to precisely time the hatching of their brood to match the exponential increase in food that spring brings.

Early April

Even with early migrants such as Wheatears, Sand Martins and Chiffchaffs already arriving in force by early April, for anyone even remotely interested in wildlife, the first Swallow sighting of the year is always a red-letter day. Having traversed the entire continent of Africa, crossed the Sahara Desert, flown straight over the Mediterranean Sea and travelled through Europe, the final obstacle – the English Channel – should present little problem to these international jet-setters. The first Swallows are usually recorded in south-west England, a mere five to six weeks after they leave South Africa, and as ever larger numbers begin to pour off the sea, they will then steadily fan out in a north to north-easterly direction across Britain. With the first Swallows often not recorded until early May in Scotland, any birds returning to northern breeding sites will be keen to make sure they're not travelling ahead of spring, which is generally considered to arrive a month later than along England's south coast. This means Scottish Swallows can afford to take their time as they feed up on the flush of invertebrates further south.

By the time the Swallows cross into Britain, most will have also completed their long, protracted moult, which will have begun as long as six months ago. At this stage the males can easily be picked out by the intense blue plumage covering their crown, mantle, rump and long, elegant tail streamers. The long outer tail-feathers of the males are much more than just a superfluous fashion statement and will soon play an important part in attracting any prospective females at the breeding sites. A higher proportion of the early birds tend to be male, with the females generally entering UK airspace a few days later. Most adult Swallows are also highly faithful to breeding sites, with those males returning ahead of the competition being in pole position to secure the best nesting sites by the time the females arrive.

As the Swallows begin to zero in on their tried and tested breeding locations, the female Tawny's patience and devotion should now finally be paying off, as chick number one wins the first of many battles by successfully managing to emerge from its egg. Despite a gap of 56 to 64 hours between when each egg was laid, in reality the chicks hatch pretty close to one another time-wise, with the chick from the final egg the only one to often emerge substantially later than the others. Tawny Owl chicks hatch in a relatively undeveloped state and initially are blind and unable to regulate their own body temperature. Arriving into the world in such a helpless state means that these scantily-clad balls of down will for some considerable time be entirely dependent on the life support system that is their parents.

The white egg tooth adhered to the top of the chicks' bill usually drops off within the first week, and during these early stages they will spend most of their time nestled under the brooding female, simply keeping warm. Initially, the chicks are unable to beg for food, but will take small morsels when offered. As their eyes are closed, it's thought they are only able to sense food when it brushes against the bristles at the base of their bill, giving the signal to 'open up'. With the female carefully tending to the chicks' every need in the nest, it will be the male's responsibility to find enough food to provision his new family. Announcing his arrival with a call, most of the prey is delivered to the nest at night and he will usually alight only long enough to drop off the food. At first their food demands are minimal, but as the chicks begin to grow and their appetites increase exponentially, their father may then also be pressed into catching food during daylight hours too.

By early April, most female Kingfishers in southern Britain will have begun to lay their first of possibly two or even three clutches deep in their subterranean nest chamber. Laying an egg a day, most clutches will consist of six or seven shiny, white and round eggs, measuring around 23mm across and weighing in at marginally more than a 1p coin. Only once the clutch is complete will incubation begin in earnest, and despite the female putting in longer shifts both parents will take their turn. When not incubating, the other bird in the pair will spend its 'downtime' either patrolling the territory or fishing.

From the moment the birds start incubating, the river tends to instantly become much quieter, but both birds will still defend their patch, and should any intruder be discovered during this period the incumbents will explode into defensive action. Exactly this type of territorial dispute was filmed by wildlife cameraman Charlie Hamilton James for his *Natural World* film entitled *My Halcyon River*. Airing on the BBC in 2002, one particular sequence has surely become one of the most memorable moments of wild behaviour ever filmed in Britain. Charlie was busy filming his local Kingfishers and managed to capture the moment that a trespassing female was caught in his resident pair's territory. Upon spotting the trespassing bird, the resident female immediately went into attack mode and, as neither bird backed down, both quickly ended up locked in mortal combat as they tried to drown each other. While this commotion was carrying on, an American Mink then suddenly dashed out of the waterside vegetation, and into Charlie's shot, before grabbing one of the Kingfishers, which was immediately dispatched and presumably stashed before the mink returned for Kingfisher number two. In the meantime, the surviving Kingfisher, which fortuitously happened to be the resident female, had managed to scrabble out of the water and to safety above ground. Even if *we* think it's a quiet time for the Kingfishers, it may in fact be anything but!

As Peregrine eggs usually take between 31 and 33 days to hatch, early April should see many urban Peregrine pairs around halfway through the incubation process. As the female carries out most of the incubating duties, the male will also need to hunt for his mate during this period. Flying in with prey, he will then either execute an aerial food pass with her or bring the item to one of her favourite perches away from the nest site. Peregrine supremo Derek Ratcliffe thought that a high proportion of kills at this time, certainly in rural nest sites, were made by the male early in the morning, and while the female fed, he would then take a turn to incubate the eggs. During the whole incubation process the only time that the eggs are believed to be left uncovered is when the female briefly leaves to preen, defecate or even hunt, if the male is either not pulling his weight or weather conditions are making catching prey difficult.

After the essential prerequisites of courtship, scrape adornment and copulation, the female Lapwing will be free to begin laying, with four eggs being the most common clutch size. Taking anywhere from four to eight days to produce her clutch, the distinctive eggs are 'pyriform' or pear-shaped, and invariably arranged so that the distinctly pointed ends face inwards. Unlike the eggs of Tawny Owls and Robins, which due to their pale, unspotted nature would stand out like a sore thumb in open country, the Lapwing's clutch has evolved to blend in with the immediate surroundings. This effort to actively camouflage the clutch is helped by the eggs' background pigment of pale brown or clay, which is then furnished with numerous black blotches, streaks and spots to break up their outline. Weighing in at around 25g each, the eggs are proportionately large for the size of the bird, meaning a clutch of four could potentially weigh close to 40% of the female's entire body weight – a huge physical investment on her part.

For Lapwings, incubation will only begin when the last egg is laid, and in common with birds as diverse as Peregrines and Kingfishers, even though both sexes are able to incubate, it's the female that takes both the larger share during the day, and the entire night session. For those females that plumped for a polygynous male, their incubating share may be even greater due to his attentions being pulled in at least one other direction.

With the nest complete, and mating largely dispensed with, the female Robin should also now be laying her clutch. Only visiting the nest for short periods at this stage, an egg will be deposited each day, with the time slot between 6am and 8am most favoured. Robin eggs are mostly white and speckled with small reddish-brown spots, although the background colour can occasionally be a pale blue or yellow. Each female Robin does seem to have inherited just one egg colour, however, with the result that all eggs in the same clutch should look similar. Like the Tawny Owl, and in fact most cavity-nesting species, the Robin's eggs do not of course need to be coloured for camouflage. Amongst British Robins, five appears to be the most common clutch size, although four, six, seven and very occasionally even more have been reported. Interestingly, clutches of six seem to be more common amongst Scottish Robins in May and June, and may be due to the longer daylight hours, giving the parents more time to forage for that extra beak. As soon as the final egg is laid, the female will then be intent on keeping a very low profile for the following 14 to 16 days of incubation, in order to avoid any unwanted attention from a whole variety of hungry predators.

Having only just built up the confidence to land at their nesting colonies, the first Puffin eggs will probably not appear for at least another three weeks yet. Initially very nervous on land, the vanguard often form 'clubs' on exposed rocks, which the birds will use as staging posts while keeping a collective eye out for predators. Spending the breeding season in a packed nesting colony means that Puffins have had to evolve a whole array of ritualised behaviours to convey messages to both their mate and immediate neighbours, one of which is the 'post-landing gesture'. This behaviour involves placing their body in a more horizontal position with their legs slightly bent and wings held aloft and above their back; this is thought to be deployed by any bird that has just flown in to appease all the other surrounding Puffins.

Unlike other colonial auks like Guillemots and Razorbills, which are primarily ledge nesters, the Puffins prefer the sanctuary of nesting underground in burrows. Upon arrival at the colony, and with their confidence boosted by their membership of the 'club', they will then start briefly peeling away in order to conduct mini inspections of any nearby burrows. This 'branching out' is conducted in the full knowledge that if any danger is sensed, they will still be able to either quickly take to the air or run back to the club. Those Puffins that have previously bred within the colony will typically first check out the very same burrow they will have used in previous years. At thriving puffinries the competition for nesting locations can be intense, with experienced birds fiercely defending what they perceive to be their own property. Any returning owner coming back to a previously occupied burrow, only to find a prospector sniffing around, will swiftly switch into full eviction mode. Puffing itself up and half spreading its wings, the aggressor will open its beak in a menacing gape, and if this doesn't have the desired effect, it will quickly pick a fight. Using its bill and feet as the main weapons, any ensuing scrap is often short and sharp, leading to the possession of the contested burrow being quickly resolved once and for all. Victory complete, the conquering Puffin will then declare his site ownership to all onlookers with a 'pelican walk', which involves standing upright, with his bill pressed to his breast feathers, while conducting a slow, exaggerated walk.

As the Puffins reacquaint themselves with their burrows, it will also be crunch time for the Cuckoos, which after feeding well in the Ivory Coast will be able to delay their crossing of the Sahara Desert no longer. Until his unfortunate demise in the summer of 2015 Chris the Cuckoo was successfully followed on his northward journey by satellite on three occasions. During those three migrations his departure point for the Sahara, and all points further north, has varied, being twice from Ivory Coast and once from adjacent Ghana. However, he always left within a small time-window of ten days – with the first of his 'desert days' falling each year between 1 April and 10 April. Chris's departure dates also seem to tally with the movements of many of the other Cuckoos tracked as part of the BTO's research programme.

Chris's migration in 2013, for example, seemed a typical year. Thought to have left on the evening of 1 April, Chris was next picked up on a juniper-covered slope, on a high

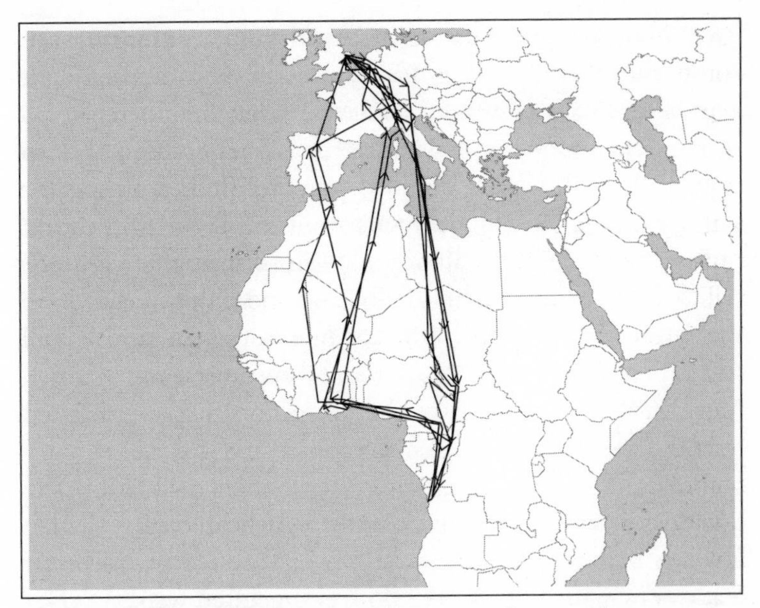

The movements of Chris the Cuckoo, showing four complete migration cycles between the autumn of 2011 and the spring of 2015.

plateau in the Atlas Mountains of Batna Province in northern Algeria. Having travelled a distance of 3,200km over both the semi-arid Sahel and then the Sahara Desert in around 60 hours, Chris was calculated to have flown at an average straight-line ground speed of 55km per hour! After a brief rest he was then thought to have moved on the night of 4 April, only to be picked up two days later, and 937km further north along the Italian and French border and close to the Mediterranean coast.

Amazingly, when Chris's route is compared with the other tracked Cuckoos emerging from the Sahara there seems a split, with some choosing their route into Europe via Italy, while others work their way towards Britain via the Iberian Peninsula. BTO researcher Chris Hewson has neatly summarised this split migration as 'two routes, one destination'. To further complicate the picture, in 2012 and 2013 Chris returned via Italy, only to then try his hand through Spain in 2014. Furthermore, when all the southbound routes leaving Britain in the summer are analysed alongside the northbound routes from Africa, it seems some Cuckoos will travel both ways through Italy, while others complete the northern and southern journeys though Spain. The most popular route of all, however, seems to be southward through Italy after breeding, only to then fly northward through Spain the following spring in a clockwise migration. And as a final variation, there seem to be Cuckoos like Chris, which appear to 'pick and mix' their routes!

Those Cuckoos choosing the western route through the Iberian Peninsula at this stage may well be close to rubbing shoulders with the British-bound Nightingales, currently still feeding away in southern Spain and Portugal. Of the five Nightingales that had geolocators attached at Orlestone Forest in Kent in 2012 and which were then subsequently recovered, all of the birds were found to have arrived back during a tiny window between 12 April and 15 April 2013. So with the final leg of their journey from southern Europe estimated to take less than 24 hours, it's a safe assumption that the birds will still be busily tucking into a Mediterranean diet right up until the middle of the month.

The Blue Tit is the last of our chosen resident birds to begin nest building, and despite the construction being carried out solely by the female, the male will often accompany her on flights to collect the material. The foundations of the nest can be just of moss, but more commonly she will start by making a basic framework of dead grass, dry twigs or rootlets. Loose moss then tends to form the bulk of the nest, which after two or three days of hard graft will have become 3 or 4cm deep. The female will then shape the cup, often in the corner of the nest, by constantly rotating her body. The final touches will include a lining of finer grasses, occasionally hair and a good number of feathers, but this last stage is only usually completed once she has begun laying her brood, which due to the large size of the clutch could take a week and a half to complete.

Early April will see the number of Bewick's Swans steadily rising across a few key estuaries and inland locations across Estonia and Latvia, as they begin to converge from other sites dotted along the Baltic coast. It seems the swans use these sites to feed on the large expanses of pondweed (*Potamogeton* species) and stoneworts, the latter being a peculiar algal group that to the untrained eye look more like underwater plants. This is obviously quite a departure from the diet of arable crops and grass pasture that many of the swans would have been consuming in Britain during winter. In fact, this choice of aquatic and marshland plants may well historically have been the swans' more traditional fare, before profound changes in agriculture here led to mass land drainage and an increase in arable land cultivation.

In a good Waxwing year, early April will still see a healthy number of birds feeding away in Britain and seemingly in no hurry to follow their colleagues back across the North Sea to their breeding grounds at more northerly latitudes. Those that have already left, however, should by now be moving on a broad front through FennoScandia, possibly following the retreating snow-melt, as they steadily work their way north to the huge taiga forests so characteristic of

Europe's northern reaches. Of course the ultimate breeding destination for these birds is anyone's guess, but of the 35 British-ringed Waxwings recovered abroad, seven came from each of Sweden, Norway, and Finland, and two from the Russian Federation. With the remaining 12 birds relocated on the near continent, and obviously actively still on migration, these recoveries give some idea of the potentially vast breeding range of this mercurial bird.

Mid-April

Making the last of their migratory hops, the Nightingale's song should be heard back in England by mid-April. Unlike many of our other summer visitors, which are usually first recorded making landfall at migration watchpoints, the first spring records of Nightingales are usually of birds pitching up straight back at their traditional nesting sites. The Nightingale is essentially a bird of dense thickets, within scrub and coppiced woodland of lowland England, and even when taking into account the worrying drop to around 4,500 singing males as recorded in the last census, the Nightingale has never been a common breeding British bird. They have never been known to breed in Ireland, Scotland or Yorkshire, for example, and despite these locations seemingly having plenty of suitable habitat, the slightly lower temperatures found further north might not be suitable for a bird that breeds at far higher densities in France and Spain.

Despite being a notoriously skulking bird throughout its entire stay in southern England, the dates of their arrival over the years have generally been easy to record as the males' first piece of business is to proclaim a territory with their beautiful, strident and ultimately ephemeral song. Singing mostly from low, dense undergrowth – although some individuals will occasionally sing out in the open – the song is instantly arresting for its immense range and power. In one study where an individual male's song was analysed, it was shown to possess as many as 250 different musical phrases, compiled from a repertoire of 600 different basic sound units. When singing, these phrases were musically arranged in an infinite variety of sequences so that each performance could be considered a unique composition – once heard, never to be repeated. Many of the phrases are rich, liquid and bubbling, but are often interspersed with guttural croaks, unmusical chuckles and dramatic pauses, the latter at least giving the human listener a chance to catch their breath.

The fact that *we* derive enjoyment from listening to the bird's song is entirely incidental, as like all bird song, it is only ever produced for the ears of its own species, having evolved as a tool to both repel competing males while simultaneously drawing in the females. Commonly, male Nightingales will arrive back a touch earlier than the females, and as they jockey for position at sites they'll be familiar with from previous seasons, it will be the 'repel' element of their song that is initially deployed.The main concentration of breeding Nightingales in Britain runs in a broad band through the counties of Suffolk, Essex, Kent and Sussex and each male holding a territory will be desperately keen to ensure that his trip all the way back from West Africa won't have been a wasted one.

While not a patch on the vocal ability of the Nightingale, with its undoubted talents lying principally in its migratory feats, the Swallow's song nevertheless also plays an important part in securing a mate for the breeding season. Like male Nightingales, the first Swallows to arrive at their traditional breeding sites will usually be the older males, who will then quickly proceed to lay claim to a breeding site of little more than a few square metres, but which crucially contains either a nest or a potential nest site. Once this miniature territory has been secured, the male will then chase away any other males showing any interest in his location and back up this ownership with song and display. His song is a not unpleasant and rapid twittering, which often ends in a harsh rattle and the subtle variations within are thought to convey important information about the bird's age (or his experience) and his health (or attractiveness). The male's tail also plays an important part in securing a site and attracting a partner, and so during these early stages of the breeding season will frequently be spread wide to reveal the striking pattern of white spots.

Once a bird that nested in caves before humans began to construct their own accommodation, many Swallows have nowadays opted for rural buildings as nest sites, with farmyard barns and cattle stalls amongst their favourite locations. The substrate to attach the mud nest to needs to be suitably rough, so cracks, corners or any type of fixing will all help secure the nest if no horizontal support, such as a beam, is available. Nest sites will also need a clear flight path for safe and rapid access and exit, and nests also tend to be sited at locations to protect the chicks from bad weather and against any variety of predators.

In terms of their breeding patterns, Swallows can be considered similar to Lapwings, in that they usually breed either solitarily or in loose, social groups. For those birds that have taken the gregarious option the territories and nests are never immediately adjacent to one another, as is the case with Sand Martins nesting cheek-by-jowl in a sandbank, but instead have a large degree of separation. For example, different Swallows will frequently hold territories in separate outhouses of the same farm, or at the very least at the opposite ends of a large barn, where an informal arrangement seems to be made which ensures that the nests are out of a direct line-of-sight with each other.

As the male Swallows begin to nail down territories in anticipation of the arrival of the females, the male Cuckoos will still be short of their final destinations in Britain. It seems the case that both the route taken back from Africa and the prevailing climate at this time of year will have a huge impact on the Cuckoos' progress as they work towards their breeding grounds. Take Chris, for example, which is still the only Cuckoo to be followed by satellite over three separate springs on his way back to Britain. In 2012, come the middle of April he was thought to be held up by poor weather in north-west Italy, while in 2013 he was in northern France close to the border with Belgium by this time, with spring 2014 seeing him in central Spain, having possibly been delayed crossing the Sahara.

With the returning Cuckoos spread in a wide European arc as they close in on Britain, mid-April at many puffinries will be the time for established couples to conduct a quick spring clean of their previously used burrows, with egg laying just a couple of weeks away. Most burrows will at some point have been constructed by the Puffins themselves, although on Skomer Island off west Wales, pairs without a burrow have been known to misappropriate Rabbit or Manx Shearwater burrows for their own use. Typically a touch longer than the human arm, the ideal burrow is usually situated on a 20–40° slope with the opening facing out to sea. This entrance will then form a narrow, blind alley dug through the soil or peat that ends in a slightly elevated and wider nest chamber. In favoured nesting sites these burrows can be at such high densities that there may only be a few centimetres of 'wall' separating some of the burrows. In a flourishing puffinry, any newly formed pairs or evicted couples may well have to start digging a new home from scratch. Excavation and restoration is carried out by both birds, with the male thought to carry out the largest share of the workload as he uses a combination of bill and feet to dig down into the peat or soil.

Compared to Guillemot and Razorbill colonies, which are deafeningly noisy, puffinries are generally much quieter places. Once members of a pair are together in their underground burrow the occasional groan can be heard, but recordings made with sensitive microphones show that in fact a whole range of vocal activity below our audible range frequently takes place just under our feet. In addition to preparing the burrow, this is also an important time for Puffin pairs both to demonstrate their bond to one another and declare site ownership. The most frequent demonstration of 'married status' to neighbouring Puffins is when a pair engage in 'billing'. This bill-clacking behaviour usually happens when a puffin lands by its mate either at a club or the burrow. This billing can quickly become an infectious activity as surrounding pairs copy the behaviour and spread the love like a Mexican wave! Individuals will also 'spot stomp' around their burrow, a ritualised behaviour that involves raising and lowering alternate feet with their webs spread, while remaining on one spot – which in the world of the Puffin simply says 'This is mine, all mine!'

As Puffins re-familiarise themselves with their burrows, it seems that the Bewick's Swans will also be reacquainting themselves with well-known and previously visited stopover sites across Estonia and Latvia during much of April. With the Russian Arctic still almost certainly frozen over in this month, it seems that the spring migration for Bewick's Swans back up north is a leisurely affair taking anywhere between eight and ten weeks. This contrasts with their autumn migration to Britain which has been calculated as only taking between four and six weeks, as the rapidly plummeting temperatures and shorter days combine to play a large part in forcing their movement towards the Gulf Stream-warmed countries of Britain and the Netherlands. During this period of the year, Eileen Rees from the Bewick's Swan research team at the WWT thinks that around three-quarters of the population of swans from the European or north-west flyway may well be widely dispersed across Estonia's wetlands. In addition to important coastal sites such as Matsalu Bay and Pärnu, where the swans will be principally feeding on pondweed, they may also be taking advantage of any ice-free floodplains and agricultural areas further inland, while waiting for the ice to retreat further north.

As Waxwings also continue their migration north with winter slowly beginning to relent, April may well see a gradual change in their diet. Having force-fed themselves on little other than berries for over six months, the first warm days of spring will see an emergence of swarms of insects – a vital source of some much-needed protein for the birds. Keen to take full advantage of any good feeding opportunities that arise along the way, the Waxwings' technique of catching insects, as they sally to and fro from open perches, will see them plucking any midges and mosquitoes straight from the air.

Back at the Tawny Owl nest, the young chicks appear to spend large parts of the day sleeping, and on the few occasions when the female needs to leave the nest to defecate or cast pellets, they will often huddle together to reduce heat loss. However, within a week of hatching, the scanty down they were born with will quickly become a far more plentiful covering. Feeding also becomes easier for the chicks once their eyes start to open after around 12 days, enabling them to take far more initiative in begging for prey items brought into the nest. Unlike the chicks of Robins and Blue Tits, for example, the small owlets don't produce their faeces in neat sacs, so their excrement is often simply eaten by the female. If the male is providing plenty of food at this stage, any uneaten remains in the nest can quickly build up, attracting flies, and so the brooding female will also have to make sure that the young chicks are regularly preened to keep them as clean and healthy as possible during this period of incarceration.

Requiring around two weeks less incubation time than the 28 to 30 days needed by the Tawny Owl, female Robins – certainly in southern Britain – will see their first chicks emerging with late April approaching. As soon as the chicks hatch, the female's first action is to remove the now redundant eggshell from the nest – part of which may be eaten if she is calcium deficient. The chicks are looked after by both parents, but the female's brief, certainly for the first few days, will be to brood them. Hatching naked, they're unable to maintain their own body temperature and will quickly die from being chilled if exposed to any April showers at this sensitive stage. During these early days, if the male brings back food while his mate is brooding he will pass it directly to her to feed the chicks. Fed on a high-protein diet of spiders, beetles and worms, the young grow incredibly quickly and will rarely spend more than two weeks in the nest before starting to fledge.

Despite laying well before the Robins, the Peregrines' lengthier incubation means that among those clutches laid around mid-March, the chicks should only now be entering their final stages of development before the egg tooth is deployed. As the female is charged with carrying out most of the brooding, the male's role will still be to provide the lion's share of food to keep both parents-to-be well fed.

Certainly in the case of urban-nesting Peregrines, feral pigeons are the most common food items consumed throughout the year, but as spring develops, Peregrine expert Ed Drewitt has found some fascinating changes in their diet. By analysing feathers collected at well-known urban nest-sites, he's been able to identify a whole variety of spring migrants caught and killed such as Wheatears, Turtle Doves, Nightjars and even the occasional Corncrake. Confined to rural farming locations in north-west Scotland, the British breeding population of the shy and retiring Corncrake is thought not to number any more than 1,200 pairs and it's patently obvious that these urban Peregrines have no respect for rarity when catching this member of the rail family during its migration north. Many of these migrants would never normally have been associated with urban habitats, but thanks to the Peregrines' opportunistic abilities, we're now learning more about their migration routes. In many cases it seems these migrant birds are flying over some towns and cities, which they may be using effectively as way-markers en route to their summer residences elsewhere.

By mid-April, most Kingfisher and Lapwing pairs will be in full incubation mode. Protected by their tunnel fortress, predation levels of Kingfisher eggs are thought to be very low, but the clutch is still rarely left unattended for any length of time to ensure the eggs don't chill. One disadvantage of nesting in almost total darkness, however, is that if an egg becomes separated from the rest of the clutch, possibly when being turned, then the adults don't seem to notice, leaving it to a chilling fate. Unable to see the outside world, the sitting Kingfisher will also have to wait for a whistle from its partner to know when a changeover is due. Upon hearing the signal that the coast is clear, the previously sitting bird will emerge blinking into the daylight, giving it an opportunity to stretch, take a toilet beak and even catch a fish or two before sitting duties recommence.

By contrast, as Lapwings nest out in the open, any changeover between parents is far more easily coordinated, and so any egg accidentally dislodged can quickly be brought back into the fold. The main problem Lapwings have during incubation is preventing their eggs being either accidentally trampled on by cattle, or predated by anything from Carrion Crows and Stoats to Foxes and Badgers. Adult Lapwings can often deter any livestock from encroaching too close to their clutch by flashing open their wings in front of the cattle, but determined predators require a more concerted and coordinated effort. Carrion Crows are considered serious predators of eggs and will be mobbed severely by both parents if they approach the nest. Any Lapwings that have bred as part of a larger colony will often be able to collectively muster an aerial bombardment, enabling them to more effectively drive the crows right out of the territory. Mammalian predators, however, are a potentially very serious proposition, as they're also capable of taking any unwary adults sitting on clutches. When faced with a Fox or Stoat, for example, the Lapwings' main reaction is to circle above the predator while calling, which not only keeps both adults safe but also warns other birds of their presence, in turn making the nest more difficult to locate. Despite these techniques, some clutches are inevitably predated and so the Lapwing's last line of defence is simply to quickly replace the lost eggs in a different nest.

Starting before the nest is finished, the female Blue Tit will proceed to lay an egg each morning until her clutch is complete. The eggs are white, with tiny reddish or chestnut speckles which are more densely concentrated at the blunter end, and weigh in at around 0.5g. The clutch size can vary enormously, between 5 and 16 eggs, although for the highest numbers there is the suspicion that maybe more than one female is laying in the same nest. Even so, Blue Tits lay some of the largest clutches of any songbirds, which fits in with their 'all your eggs in one basket' strategy. Younger females tend to lay smaller clutches, while those birds nesting in woodland often lay more eggs than those who have chosen farmland localities, who in turn will frequently have higher broods than any Blue Tits which plumped for a garden nest box. Despite gardens being incredibly important for keeping hungry birds fed during the cold winter months, when it comes to the breeding season, it seems that deciduous woodlands, with their huge spring concentrations of caterpillars, will always trump anything that any other habitat has to offer.

Late April

After around four weeks of constant incubation, the female Lapwing's hard work will finally pay off as her clutch begins to hatch. Using their strong neck muscles and egg tooth, the chicks will break out by both bracing themselves against the inside of the egg while tapping away from inside to crack the exterior. Once the Lapwing chicks have recovered from their efforts and dried out sufficiently, their already advanced state of development means they're able to run around and feed themselves almost immediately. The precocious chicks are delightful little creatures, with long blue-grey or pinkish-grey legs, a short dark bill and sandy-brown down, which is dotted and streaked black. They also have a very distinct white patch on the back of their neck, which stands out when the chicks are running around and almost certainly enables their parents to easily keep track of them. However, the instant they feel threatened, or in response to their parents' alarm calls, they will crouch motionless with their legs drawn in and heads down in order to hide the white patch and melt into their surroundings. Not needed to find food for their young charges, the parents' role is both to lead them to the good feeding areas and guard them from any marauding predators eager for an easy meal.

A mosaic of habitats seems to suit the chicks best, with wet patches and short swards for feeding, while also comprising longer, more rank vegetation for concealment should danger appear. Ideally these good feeding areas will be present within the parents' territory, but if not, the adults will quickly corral their offspring away to other locations, possibly hundreds of metres away. As the chicks are unable to fly so soon after hatching, any impassable obstacles encountered on the way can usually be negotiated by the parents simply airlifting their chicks over to the other side. The feeding behaviour of the chicks seems most closely to resemble that of back-garden chickens, with the chicks scratching around the central figure of their mother while they search out beetles, larvae and earthworms. They also feed like their parents, by moving and scanning, with any chick straying too far instantly called back to the fold. Despite being able to find more than enough food themselves, they can initially struggle to keep warm during poor weather so will still disappear under their mother to be brooded on cold, wet days and each night.

For the few days before they eventually break free of their ovoid prison, the incarcerated Peregrine chicks will begin cheeping to their parents, who often respond, before becoming noticeably more restless as they prepare for the youngsters' arrival. The chicks will usually emerge within the space of a couple of days, although not all eggs will necessarily hatch, as a small proportion tend to be either infertile or suffer from the embryo having died during incubation. After their exertions to break free the young will initially be weak, wet and exhausted, but then quickly dry out into fluffy, white chicks. Unlike the Lapwing youngsters, the Peregrine chicks hatch in a far less advanced (or altricial) state and so are utterly reliant on their parents for some months before they are able to catch their own food. Upon hatching, the parents don't bother to remove the eggshells, which instead are trampled into the nest by the chicks and parents over time.

The appearance of the chicks will result in the parents suddenly turning much more aggressive and intolerant of trespassers, driving any non-related Peregrines and potential predators away from the nest site. With just their first coat of down to protect them from the elements the young can be incredibly prone to chilling, and so will initially be brooded almost continually by the female. Huddling together to conserve heat, the young are then fed tiny morsels of food by their mother as they take the first of many meaty meals caught and brought in by their diligent father. Even when the chicks are small it won't take long for a hierarchy to develop within the brood, with some chicks always demanding to be fed first, while any weaker chicks are forced to feed afterwards. In lean times, small chicks can often quickly disappear from the nest, but in years of plenty the entire brood may well stand a good chance of surviving, certainly as far as fledging.

Also making an appearance before the start of May, the Kingfisher chicks, like those of the Peregrine, tend to emerge at the helpless end of the scale. Hatching into their dark, dingy nest chamber, the chicks are pink, blind, devoid of feathers and will be totally dependent on their parents. With usually just a few hours between the hatching of each chick, the adults' first act is to remove the broken eggshells, presumably to maximise the space in the nest chamber. The chicks will often take their first fish meal within just a few hours of hatching, and as the prey items are not dismembered by the adults, but eaten whole and head first, the parents will initially bring in quite small fish of around 5cm or less. Sometimes it can look like particularly young chicks are struggling to swallow a big fish, but their guts are very effective at quickly breaking down the food to make room for the rest of the meal. Unable to see for at least the first eight days, it seems that during this period the young respond to the light being blocked out from the tunnel as the sign that a parent's arrival with food is imminent.

For those Tawny Owl chicks hatching in early April, by the end of the month their eyes should be open and their bodies completely covered in a grey jacket of down, far more adept at keeping them warm than the sparse covering with which they hatched. As the chicks become more able to insulate themselves against the vagaries of the British spring, this should in turn free up the female to help her mate find enough food to supply the brood's ever-increasing appetite. Despite now spending less time with the chicks, the female rarely travels far, as she leaves her partner responsible for foraging further afield. When food is brought in by either parent, the nestlings will by now be jostling with each other for the best position to take any food offered. This undignified scramble also involves much noisy begging and there is some evidence that the adults tend to feed the chick calling the loudest. Any chick that has secured a good prey item will also call vociferously immediately afterwards, and this could help it defend any meal too large to swallow instantly, as at this stage the youngsters are still too small to tear it up. Even when the adults are not present, the chicks will still call frequently, possibly in order to establish a pecking order, which could mean all the difference between life and death in those years when the demand for prey outstrips supply.

Faced with having to find ever more food for their growing chicks, the Tawny Owl parents may even be forced to hunt during the day just to keep up with the incessant demand. Tawny Owl expert Dave Culley has carefully studied the diet of his pair of Tawnies in suburban Cheshire and found it far more varied than just mice and voles. The chicks in his nest were fed a diet of around 50% small mammals in late April, but as other young birds, such as Blackbirds and Robins, fledge from surrounding nests, songbirds soon become the major share of the young owls' diet. A couple of weeks after hatching will also see their wing and tail feathers emerging from the mass of down, and although skeletal growth of the chicks generally finishes much earlier, the flight feathers will continue to grow even after they have left the nest. At a fortnight old, the chicks are now looking far more 'owl-like', and will begin to exercise, preen and scratch much more. Head bobbing can also be seen – a technique used to gauge distance, and a skill they will need to thoroughly master before they begin hunting for themselves later in the year.

Three days after hatching, and providing the Robin chicks remain well fed, the appearance of quills all over their skin indicates that feather emergence should quickly follow. A couple of days further on will see the chicks' eyes also begin to open, and such is their rapid rate of development that by the tenth day their bodies will be more or less fully feathered. With the Robins at the point of fledging less than two weeks after hatching, a full brood will see the nest literally bursting at the seams. Still devoid of the famous red breast until after their first moult, the heavily speckled youngsters will by now have learnt the art of crouching down and flattening themselves if danger threatens. But as soon as the coast becomes clear, they will instantly change their tune back to that of noisy beggars, with a sea of yellow gapes and cacophony of calls confronting any parent turning up with food.

Certainly in southern Britain, as April draws to a close most female Blue Tits should have laid close to their entire clutch, with incubating generally beginning just before the final egg is laid. As laying takes place, the female sheds feathers from her breast and belly to expose a large area of skin that becomes exposed, wrinkled and almost purple due to the rich supply of blood vessels running just below the surface. When this brood patch is combined with the insulating properties of the nest and feathers from around the bird's breast, which operate as a skirt to seal in the heat, this should easily provide sufficient warmth to incubate the eggs at the optimum temperature for their development.

Only too aware that the time for synchronised egg laying will be rapidly approaching, from securing a burrow and carrying out repairs, the male Puffin will turn his attentions to consummating the relationship with his mate. For established pairs, Puffin courtship may have begun even before the birds make their spring landfall, with most matings tending to take place on water. Invariably instigated by the male as the pair bob around like corks on the surface, he will make his intentions perfectly clear by flicking his head back amid a fluttering of wings. If the female is not interested she will constantly dive to keep him at bay, but if he catches her in a more receptive mood, swimming low in the water will encourage the male to mount her as he uses his wings to balance. The coupling is once again a very brief affair, usually lasting less than 30 seconds, and afterwards the male will usually guard his female to ensure he is not cuckolded before his mate lays their single, precious egg in early May.

Arriving later than the males, most female Swallows will generally find a partner within three days of returning to their traditional nesting sites. However, the 'fairer sex' can still be quite choosy and the male will need to use all his persuasive powers if a female is to select him for the breeding season. A good nesting location will play a large part in attracting a female, as will the male's looks, with Swallow researcher Angela Turner stating that 'in Europe at least, males with longer tails are clearly the choice of discerning females'. The fussy females also seem to additionally prefer males with outer tail-feathers of the same length, those individuals with larger spots on their tail feathers and any with particularly bright red coloration in the face. In essence it seems in the world of the Swallow that long tails and bright colours are an indication of good health and also possibly more experience, as more sickly males will not have fed sufficiently well for the extravagant feather growth.

Reaching 'first base' by securing a female's attention, the male will then make repeated flights up to his nest in an attempt to entice her into admiring his residence. If she is then enticed over for a closer inspection the male's next move will be to land close by in order to show off his fanned tail, while uttering notes of encouragement, as he attempts to seal the deal. In locations where a number of pairs may be nesting in a loose social flock, some particularly handsome and persuasive males may be able to secure two females to mate with, while short-tailed males may attract far less interest. For those males lucky enough to secure at least one mate, the next job will be to either repair an old nest – should one still be present in a reasonable state – or alternatively start building a new one from scratch.

As the female Nightingales arrive back, they should already find their breeding grounds carved up into territories as the males implore the females to visit their own real estate. The male Nightingale is of course famous for singing both day and night, but exactly when their song is produced conveys a different meaning. The diurnal song is mainly directed at approaching rivals and associated with holding territory, while the nocturnal song is thought to be more of a long-distance advertisement to any females looking for a partner. During the day the males sing from several perches and regularly change position with the start of each song – this contrasts with the nocturnal song, which is usually delivered from one particular perch that may be favoured for several nights in succession.

Although physical fights between males do occasionally happen, the main form of defence and attack is the territorial song. Adjacent males intent on courting a visiting female will often have long song-duels at the territory border, which may then occasionally escalate to a bowing threat posture. If this aggressive and overt display has still not persuaded either male to back down, they may then chase each other through the trees, singing as they go, until a winner emerges. Once the female has made her choice, the pair will then undergo a ritualised pair bonding where the male hops around on the ground below the female as he pours forth his finest repertoire. During the delivery of this virtuoso performance he will often then seal the deal by hopping up to her branch while raising and fanning his tail to top off the display. With the bond cemented and the relationship duly consummated, the male will then continue to converse with his mate around the territory through short and fragmented bursts of song. With the nocturnal song having thus achieved its main aim, this slowly begins to tail off as the pair prepare for the next phase of the breeding season.

With Cuckoos still dotted at various sites across Europe, late April may well be their last chance to feed up before the final leg of their migration sees them converging on the English Channel as they prepare to cross into Britain. Depending on the weather some Cuckoos can be heard calling in Britain in April, but the data from all the Cuckoos transmitting their position to satellites as part of the BTO's research suggests there is remarkable uniformity, with the first week of May being the time when most males seem to arrive back. In the four springs that Chris the Cuckoo successfully navigated his way back to Britain, his arrival dates were 1 May in 2012, 4 May in 2013, 25 April in 2014 and 27 April in 2015.

Remarkably, while many Cuckoos are within a hair's breadth of reaching Britain, in those springs following a 'Waxing' winter there may still be a small number of birds dotted about the British countryside yet to leave. With no confirmed breeding records of Waxwings in Britain, these late departees have presumably decided that staying and feeding in Britain with a view to returning rapidly to their breeding grounds is a more favourable tactic than leaving earlier in the year and taking the slow train. For those early departees taking the leisurely continental route, late April should see them closing in on their summer destination. Their approach to the breeding grounds is believed to be the time when pair bonds form, primarily because there seems little evidence of courting birds amongst the overwintering flocks in Britain.

Still in no immediate hurry to reach their breeding grounds, most Bewick's Swans that overwintered in Britain will still be busily foraging at a variety of sites across Estonia, the Gulf of Finland to the north and even the most western parts of the Russian Federation. Keen to feed as well as possible while waiting for the big thaw to reach the maritime tundra further north, the swans will need to have put down plenty of reserves in preparation for heading north-east across the vast forest tracts of the Republic of Karelia, to the appropriately named White Sea and all points beyond.

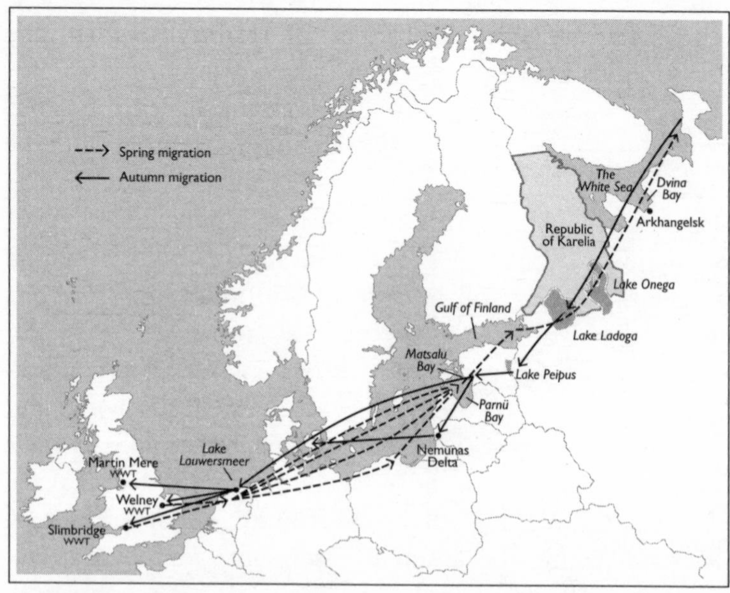

The believed spring and autumn migratory routes of Bewick's Swans between their wintering grounds in Britain and northern Europe.

May

For the naturalist, May is simply a fabulous month for watching wildlife, making it difficult to know where to turn. As the leaves on the trees unfurl, our woodlands take on a vivid green hue, which contrasts with rural hedgerows being painted white as Hawthorn, or May Flower, comes into blossom, while the yellow blaze of Gorse lights up our moorland and heathland. During this month of 'May-hem', Badger and Fox cubs will be emerging from their underground setts and earths, while any mammals which spent the winter in hibernation will be playing catch-up as they get stuck in to the fighting, feeding and breeding season. The dawn chorus at this time is deafening, as all the migrants compete to make their voices heard above our resident birds. The warmer ambient temperatures this month will also produce a super-abundance of invertebrates, on which parents with hungry chicks will be able to capitalise, and with the sun rising steadily earlier and setting later they shouldn't be short of daylight in which to forage either. Away from our shores, those birds that spent the winter in Britain should also be completing their migration as they reach distant breeding grounds spread right across both northern Europe and the Arctic regions of the Russian Federation.

Early May

The 'advance guard' may have already arrived during the last few days of April, but early May has to be the time to listen out for the first Cuckoos arriving back after a winter spent in the hot, humid forests of Central Africa. Surely one of the most recognisable calls in Britain, Wordsworth's 'wandering voice' seems the very essence of spring, and yet is produced by a bird that is seen by surprisingly few people. Breeding throughout much of Britain, Cuckoos can be found in an array of habitats, with the exception of built-up areas, but most commonly favour reedbeds, moorland, woodland and agricultural land. Certainly amongst British birds, the Cuckoo is a bird with a unique lifestyle, being the only British species to never build a nest but preferring instead to lay its eggs in the nests of smaller foster birds, or hosts, which incubate and rear the young to their own detriment – behaviour that is called 'obligate brood parasitism'.

The distinctive '*cuck-oo*' call is only produced by the males, who usually arrive a week earlier than the females, and start advertising their presence as soon as they turn up at their respective breeding grounds. Rather than holding a territory in the conventional way, the males tend to have a 'song range', which can vary in size but tends to average around 30 hectares. This 'range' may well overlap with other competing males, but also crucially will encompass the 'breeding range' of at least one female, when they arrive back a touch later. The famous disyllabic call can sometimes be monotonously repeated up to

300 times in succession and is capable of carrying up to 5km in favourable conditions. Anyone who has ever tried to observe a calling Cuckoo knows their voice has a ventriloquial quality, which is a result of the bird turning its head while calling. Some naturalists with a good musical ear claim to be able to separate individual birds on call by subtle variations in pitch and tempo, and so it seems highly likely that the male and female Cuckoos must also be able to recognise each other's calls.

The Cuckoo is also the only British bird known to actively seek out the seemingly unpalatable hairy and brightly coloured caterpillars which form such an important component its diet. But this food supply is rarely abundant early in the season and observations seem to suggest that the Cuckoos substitute the caterpillars with a whole variety of beetles, spiders, ants, flies and worms until their favoured prey item comes on tap. The Cuckoo's main feeding technique involves scanning the surrounding environment from a perch. When the movement of a suitable item is spotted the Cuckoo will fly down to make a capture, either then demolishing the prey on the spot or returning to a perch to deal with the catch. Before being eaten, any large caterpillar is worked through the Cuckoo's bill from end to end, prior to being given a quick, violent shake to ensure the larva's gut contents are ejected before the eviscerated item is then swallowed. The caterpillars' hairs collect in the Cuckoo's stomach lining, which is then periodically shed before being regurgitated as a pellet.

Female Cuckoos seem to be incredibly site-faithful, with the vast majority returning to the same location year after year. Each female Cuckoo has evolved to specialise on one particular host species, which will be victimised throughout the lifetime of the bird. Although female Cuckoos have been recorded laying their eggs in the nests of over 100 different species of birds according to the BTO, just five hosts account for 90% of all parasitised nests. In marshland Reed Warbler is the main host, and in moorland and heathland, Cuckoos opt for Meadow Pipits; Dunnocks or Robins are mostly targeted in woodland and farmland, while the Pied Wagtail is often selected in open country. This pre-ordained selectivity divides the Cuckoo into several host-specific races, or 'gentes' - a term coined by the famous Cambridge zoologist, Alfred Newton.

On arrival back at their traditional site the females will immediately attempt to take over a distinct breeding range of around 30 hectares, which will commonly hold a healthy population of her host species, and into which she will plan to lay all her eggs. Often this appears to be an exclusive territory, but at some locations it can contain a number of females with overlapping breeding ranges. When there is some overlap, often just one female will dominate proceedings and lay far more eggs than the other subordinate birds. Early in the season any newly arrived females will often be harried by amorous males, and so they will prefer to a low profile until they're ready to mate.

Early May should also see the arrival of the first Waxwings back at their breeding sites in the higher latitudes of northern Europe. Called 'boreal' in Canada, or 'taiga' in Russia, the huge tract of forest that covers large parts of Canada and Alaska, most of Sweden, Finland and Norway, and much of the northern Russian Federation east to Japan is considered the world's largest land-based biome. Representing an astonishing 29% of the world's forest cover, taiga is relatively low on biodiversity, when compared to, for example, the tropical forests, primarily due to the astonishingly low winter temperatures. With a record low of −53°C in the Russian taiga, many mammals will hibernate to avoid the worst excesses of the winter, while most birds will simply migrate to warmer and lower latitudes. But by July, the temperatures will often rise well above 10°C, and when this is combined with the long hours of daylight and scarce competition for the food available, the habitat should give hardy birds like

Waxwings more than enough time to rear their broods before autumn brings the short breeding season rapidly to a close.

It's generally assumed that most Waxwings overwintering in Britain will be birds from the taiga forests of northern Norway, Finland, Sweden and the north-western parts of the Russian Federation, which made one particular ringing recovery all the more astonishing. One Waxwing ringed as an adult male in Aberdeen on 31 March 2005 (during the large invasion of the winter of 2004/05), was subsequently reported to have been killed by a cat the following February in a village located in the Khanty-Mansi region of the Russian Federation, east of the Urals and north of Kazakhstan – or 3,714km from its British ringing location! This spectacular long-distance movement aside, of the 35 Waxwings ringed in Britain and reported elsewhere, 21 have been recovered from Norway, Sweden and Finland, with a further 12 thought to be actively migrating birds recorded from the near continent. Irrespective of the line of longitude to which the Waxwings return, they will always nest south of the latitudinal line where the trees finally give way to Arctic tundra. The taiga forest itself consists primarily of huge stands of Spruce and Pine, mixed in with Birch and plenty of fruit-bearing plants such as Bilberry, Cloudberry and Cowberry. Lichens are also abundant in these forests, and the hanging Witch-hair Lichen (*Usnea*) festoons many of the conifers in which the Waxwings will ultimately nest come the middle of June.

Deciding they too must head north, early May should see the majority of Bewick's Swans on the next leg of their epic voyage, which will see them departing to the White Sea on Russia's northern coast. Leaving from a wide variety of sites across Estonia, Latvia and the western outreaches of the Russian Federation, the swans will fly in a north-easterly direction, possibly over St Petersburg and into the Republic of Karelia, a sparsely populated land composed primarily of trees and lakes. This federal subject of the Russian Federation has a long western border with Finland, is demarcated to the north by a 630km coastline with the White Sea and to the south with a convoluted border which abuts Europe's two largest lakes, Lake Ladoga and Lake Onega, themselves situated within mother Russia. It's thought that the swans will either stop temporarily at these lakes or use them as way-markers while they take a north-easterly bearing, the direction which also represents the shortest overland distance to the White Sea. The brackish waters of the huge Dvina Bay and Dvina Delta, situated alongside the major Russian port of Arkhangelsk and within the White Sea, are around 1,000km from Estonia and represent an internationally renowned staging post for Bewick's Swans, both on their spring and autumn migrations. With many having flown directly from the Baltic, the swans are then thought to feed on the eelgrass and stoneworts in the bay while waiting for the sea ice to recede. This area is considered so important that in spring 1996 18,000 Bewick's Swans were estimated to have passed through the Dvina Bay region, a figure which at the time may have represented over 60% of the entire population of Bewick's Swans from the European flyway. Upon arrival the swans are thought to rest and feed up for a short time before then continuing to filter in a northwesterly direction which also follows roughly along the line of the Russian Federation's coastline. Passing through the strait that separates the White Sea from the Barents sea, the swans should be aware they're on the home straight as they press on to the maritime tundra further along the coast.

Even before the Bewick's Swans have arrived at their breeding grounds, early May back in Britain should see many Robins' first broods taking their first leap into the big, wide world. Leaving the nest a mere fortnight after they hatched, the Robin chicks' flight feathers will still be growing when they take their leap of faith, so the fledglings will still need a couple more days until they're actually able to take to the air. At this stage many first broods will be abandoned by the female as she leaves them to be looked after solely by their father, in order to turn her attention to building another nest for her second brood. Despite dad's best efforts, the first few days outside the nest can be brutal for an inexperienced, flightless Robin chick, with many simply starving or falling prey to a whole raft of predators. Heavily speckled to disrupt their outline and aid camouflage in the undergrowth, those chicks that have managed to survive their first *septimana horribilis* should slowly begin finding at least some of their own food.

With the Lapwing chicks also vulnerable to predation at this early stage, their father will assume the mantle of chief lookout and guard. Standing slightly away from the mother and chicks, and in a good location to spot any potential predators approaching his charges, he will harry any perceived threat until they are driven away. Stoats or Weasels will be mercilessly mobbed, dogs can be attacked and he will often circle above humans while alarm calling, in an attempt both to distract the attacker and give the beautifully camouflaged chicks a moment to crouch down and freeze. As is the case for many young chicks, mortality is highest in the first few days, with one Dutch Lapwing researcher reporting that despite the best effort of the parents, 60 to 75% of all chicks were lost to predators within the first ten days. After this unrelenting period the number of chicks taken seems to steadily decline as the young not only grow stronger, but also become more streetwise.

Able now to feed without needing their mother's help in dismembering prey, the young Tawny Owl chicks will soon become keen to break free of the constraints of nest life too. In most years, early May should see them spending long periods of time at the nest entrance before finally pluking up the courage to make a bid for freedom a few days later. The oldest chicks are evidently the first to leave the nest and will often be encouraged out by the soft contact calls of their mother from a nearby branch. Unable as yet to fly, their mode of transport will be climbing as they use a combination of talons, bill and flapping wings to work their way up into the branches of the tree – a process unsurprisingly called 'branching'. At this early stage many chicks will fall out of trees, only to end up on the ground, but they innately seem to understand this is not a safe place to spend any time and so are usually able to quickly scramble back up again. After having left the nest site the young will not return again, but rarely travel far and often spend the first few days huddled together as both parents continue to bring in sufficient food to meet their insatiable demands. Still a mass of pearl-grey down, with their wing feathers bursting through, the chicks will spend at least the first week out of the nest precariously perched amongst the branches before they build up enough courage to take their first, silent flight. This maiden voyage will often be nothing more grandiose than a short gravity-defying flap between two trees, and will certainly be a case of one small step for the Owl but a giant step for Tawny-kind.

Still a full month from fledging, the Peregrine chicks will nevertheless be putting on weight and becoming stronger with each passing day. Even at this tender age, preening, scratching and wing or leg stretching all become part of the daily routine, in addition to their favourite occupation of sleeping. As they begin to grow a second, warmer layer of down at around ten days the female will need to brood them less, meaning that unless the weather is wet, cold and windy she will begin leaving the nest for short periods. As their vision starts developing strongly, this will lead to the youngsters suddenly becoming more mobile. Able to observe any comings and goings from their lofty position will often result in a mad, noisy scramble towards any parent spotted arriving with food. At this stage the male will still be catching the majority of the prey for the chicks, but he will from now on also begin helping the female to feed the chicks at the nest. Arriving at the nest site, the adult dishing out lunch will pin down the plucked and headless prey with its powerful talons, before proceeding to rip it up and feed one morsel at a time to the chicks. As the young become increasingly more coordinated, noisy and hungry, the chunks of food they are able to deal with will increase in size.

In their tunnel nest, little will be heard from young Kingfisher chicks until their eyes start opening at around eight days. The power of sight will suddenly shake them into turning the volume up, leading to their constant churring calls frequently audible from the bank side, particularly when they're hungry and waiting to be fed. Even before the first feathers break through their translucent skin, the blue colouring so characteristic of Kingfishers is very noticeable, and the imminent arrival of plumage is timed to match a huge increase in the chicks' demand for fresh fish deliveries. As each chick is fed its fish it will then shuffle to the back of the pack to digest the meal and await its next turn. Any chick daring to get more than its fair share by queue jumping will often be pecked back into line by its hungrier siblings.

In those nests where the parents are able to catch enough fish, renowned Kingfisher expert David Boag calculated that each chick will be fed around every 50 minutes, so a typical nest could easily consume over 100 fish a day at its peak, and this figure doesn't include those consumed by the adults themselves. By now the chamber will be reeking of regurgitated fish remains from both the chicks and their parents, but as the youngsters become more mobile they seem to be aware of this sanitation issue and so begin venturing down the tunnel to defecate rather than add to the mess in the chamber. Squirting out their white poo soon turns the tunnel into a slimy mess through which the adults must trudge through each time they return with food. This necessitates frequent bathing and preening sessions for the parents to ensure their plumage remains both flightworthy and waterproof. The sheaths that enclose the feathers will also begin poking through the skin at this stage, leading to the chicks quickly turning into a spiky mass of blue and orange. Interestingly, the feathers remain encased until they are virtually full grown, with the sheaths only dropping off shortly before fledging. This is thought to be an adaptation to protect the new feathers from the filth and squalor of the chamber, to ensure the plumage remains in the best possible condition for the moment the youngsters fledge.

With her clutch complete, early May should find many southern Blue Tits in full incubation mode. The sole responsibility of keeping the eggs warm and turned on a regular basis falls to the female, who will spend the entire fortnight ensconced in her cavity nest, bar a few short breaks to feed, drink and toilet herself. During this vigil she will sit tight, and as many generations of young egg collectors and bird ringers can attest, the nickname 'Billy Biter' is well deserved. Despite her devotion, it's thought as many as a third of all Blue Tit clutches are predated by Wood Mice, Great Spotted Woodpeckers and Grey Squirrels. While the female incubates, the male has relatively little to keep him busy apart from feeding himself, and with his paternity already assured he will not feel the need to regularly sing during this period. He will, however, remain in sight of the nest entrance for the duration, and often escort the female on the occasions she briefly needs to leave the nest.

In contrast to the large Blue Tit clutch, no pair of Puffins has ever been recorded laying anything other than just a single egg. Despite a difference between years and a natural variation between colonies, early May, in most years, will be the peak egg laying period. The whitish-coloured egg, weighs in at over 60g or around 15% of her body weight. Unlike the eggs of Guillemots and Razorbills, which are pear-shaped to help prevent them being accidentally knocked or blown off their precarious nesting ledges, the Puffins' underground egg need only be elliptical. After it has been laid, both members of the pair will take it in turn to incubate the egg, which is tucked against the bare skin of one of the brood patches located underneath each wing. Incubating in shifts, each stint will often last more than a day while the other partner feeds out at sea. Taking an average of 41 days, Puffin incubation is such a time-consuming affair that it won't be until mid-June that the first Puffins will be seen coming ashore with fish, a sure-fire sign that their egg has transformed into a newly hatched underground 'puffling'.

Close on the heels of the Puffins, those Nightingale pairs already established will begin nest-building in early May. The female, accompanied by the male, tends to select the precise spot to rear the young which is usually located on or near the ground, surrounded by dense vegetation and often underneath some shrubby growth. The construction duties are also her responsibility, as she starts by moulding a loose, bulky cup of dead leaves and grass, before then lining it with finer grasses and hair. The nest complete, she will begin laying her clutch of olive-green to olive-brown lightly speckled eggs, only commencing incubation when the last of the four or five eggs is produced. As the whole incubation process rarely takes more than 13 or 14 days, this means productive Nightingale sites should be awash with hungry baby chicks well before the end of the month.

For any Swallow pairs choosing to build a nest from scratch the assembly period may take around a week, and sometimes even longer when unseasonably cold or dry weather makes suitable mud harder to find. Variable amounts of dry grass, straw and even horsehair are all added to strengthen the cup, which is principally made up of over 1,000 pellets of mud that have been collected from a variety of suitable locations nearby. Built by both birds, the first stage of the process tends to be the construction of a ledge on which the birds can perch, and from this base they can then build the walls upwards and outwards. Only spending a few seconds on the ground as they collect mud from a nearby stream's edge or hoof-print of a domestic animal, the mud is pushed into place using the bill and tongue, with the Swallow additionally vibrating its head up and down to ensure it is tamped down into place.

Most building takes place in the morning, which will allow the mud to dry well before nightfall and also free up the pair to feed during the warmest part of the day when insects will be most abundant. Once the external structure is around 20cm wide and 10cm high, comprising a cup deep enough to hold the chicks, the pair will trun their attentions to the interior soft furnishings. Lining the cup with dry grass, soft hair and white chicken feathers, no doubt collected from around the farmyard, the feathers in particular are much more than just a fashion accessory and will help to insulate the nest during incubation and prevent the eggs chilling too quickly in the brief moments that the female must leave for a stretch and a feed.

Mid-May

Totally absent during the egg-laying and rearing process, the male Cuckoo's success must be measured not just by the number of matings with females, but also by his ability to keep any other competing males at bay in order to ensure his paternity. Upon arriving back at their regular breeding sites, the females will initially be keen to avoid the males, as more pressing issues will need to be addressed first, such as feeding up, and carrying out an assessment as to the state of the host nests in which they will ultimately hope to deposit their eggs. As it appears the best places for host nests are not always the best locations for caterpillars, most Cuckoos are thought to forage away from their breeding ranges. This means both sexes will have to periodically and temporarily suspend any breeding activities while away feeding in nearby locations, such as orchards with a rich supply of caterpillars.

While needing to be in the best possible condition to produce her eggs, some females must be careful not to delay the breeding process any more than is absolutely necessary, as late arrival to the party may have already resulted in them missing opportunities to parasitise the first clutches of resident breeding birds such as Dunnocks, Robins, Meadow Pipits and Pied Wagtails. However, for those Cuckoos that have evolved to specifically victimise Reed Warblers, the need to lay quickly may be less of an issue, as Reed Warblers, like the Cuckoos, are summer visitors, meaning they will invariably be laying their first clutch later than the resident species.

Once the female Cuckoo's initial assessment of the state of the surrounding host nests has been completed, she will then be in a position to succumb to the male's overtures. Picking up on this sudden interest, the male will respond by singing excitedly, while displaying with a bowed posture, his wings drooped and his tail held, half-erect and fanned, to expose the white tips and notches. He may also rotate his body and swivel the fanned tail, as the female quietly watches – no doubt assessing his calibre. Astonishingly for a bird that doesn't make a nest, the male will often fly down and even offer her a piece of grass, leaf or twig, but courtship feeding is only believed to occur very occasionally. After the relationship is consummated, the female's next job is to continue her watching brief, as she quietly sits in any bushes or trees that afford discreet views of any host nests she may have already earmarked for an egg.

The fact that each female Cuckoo specialises on a certain host species was first discovered by the pioneering work of Edgar Chance on a Worcestershire common between 1918 and 1925. Chance noted that each Cuckoo egg tended to look remarkably similar to those of its chosen host species, and only ever appeared during the host's laying period and, crucially, before the completed clutch meant incubation had already begun. As each host is deemed suitable for parasitism for such a small window of time, consisting of no more than three or four days, Chance realised that the female Cuckoo must watch her victims very closely in order to get her timings right. He subsequently discovered that the eggs are laid into different host nests at two-day intervals, with the crime always perpetrated in the afternoon or early evening. This interval not only gives the female Cuckoo plenty of time to form another egg but also to locate her next nest, enabling the egg to be quickly introduced at the precise moment of her choosing, when the parent hosts are least likely to be present at the nest. Ideally the Cuckoo will try to lay into those nests containing only one or two eggs, to not only minimise the chance of her own egg being rejected by the host, but also to ensure that her chick hatches as early as possible, in turn making the Cuckoo chick's job of ejecting the host's own eggs or chicks much easier.

Cuckoo expert Nick Davies from Cambridge University has watched Cuckoos parasitise Reed Warbler nests on a number of occasions: 'Before laying, the female remains motionless and hidden in a tree, up to 100 metres away from the host nest. Then, after a period of anywhere between 30 minutes and two and a half hours, she glides down to the nest and lands on the rim before bowing into the cup; a few seconds later she lifts her head, holding one of the warbler's eggs in her bill, and then moves to sit on the nest. Her abdomen moves down as she lays her own egg, then without even a single glance into the nest, she flies off, still carrying the warbler's egg in her bill. She lands in the bushes 30 metres away, swallows the egg whole and then utters a strange bubbling call as if in triumph.' Both Nick Davies and Edgar Chance reported that in each case the female usually conducted the whole process of stealing a host egg and then replacing it with one of her own in less than ten seconds.

The total number of eggs the female is able to lay in a season will be limited by the number and state of host nests available, but some Cuckoos are able to maximise the number of eggs they can deposit by manipulating their hosts. By predating any clutches that are at too advanced a stage for parasitism, thereby causing the hosts to re-lay, this should enable her to take advantage of these nests the second time round. By manipulating the Meadow Pipit clutches himself, Edgar Chance observed one female laying an incredible 25 eggs in a single breeding season. However, this was equalled, without human interference, by a female Cuckoo studied by ornithologist Mike Bayliss in Oxfordshire in 1988, who recorded one female working over a population of 36 Reed Warbler nests, resulting in 24 pairs being successfully parasitised, including one poor pair twice! These records aside, it's thought more likely that the number of nests parasitised by most females will be around eight, with subordinate females at some locations often prevented from laying any more than just a couple of eggs.

Laying her clutch in the far more conventional manner, most female Nightingales should be sitting on a full clutch by the middle of May. Even though the male will occasionally feed his mate while she incubates, the pair are thought to find their own food, with the male using his courtship song to summon her off the nest as and when the coast is clear. However, should the male spot a potential predator, he will also give her a warning whistle to sit tight in the nest. Hoping not to be spotted is her best form of defence, but she will never sacrifice her own well-being just for the sake of a clutch, and so will always slip away at the last moment if she feels in mortal danger herself. The male tends to sing less during the laying period, but will then strike up his song briefly again as his partner begins to sit. Due to the warmer summers in continental Europe, many of the Nightingales are double-brooded, meaning the male will need to hold his territory for much longer, but as English Nightingales are right at the edge of the species' range, it is almost certain that our birds will only ever rear one brood. One unfortunate knock-on effect of the shorter British breeding season is that with the male's beautiful song having already fulfilled its purpose it will be heard far less often as the month of June progresses.

As most Puffin burrows are either on islands or at sites that rats have been unable to colonise, such as Bempton Cliffs

RSPB Reserve in East Yorkshire, predation of any of the subterranean eggs is considered to be minimal at most puffinries. The reasonably constant temperature underground means the sitting bird shouldn't need to warm the egg continuously, leaving them free to take short breaks above ground for a preen and a stretch. Changeovers between the parents generally occur in the morning or evening, with the period just before dusk also seemingly the key time when most Puffins socialise above ground with their peers. These 'socials' are thought to fulfil an important function within the colony, and are used by each pair to maintain close and regular contact with their neighbours. When both parents are present during this fraternising hour, their agenda will include checking up on the progress of pairs close by and peering down adjacent burrows. As night falls, the slopes will become deserted, as the birds either fly out to roost on the sea or retire underground for a night's egg-warming.

For those Swallows that have chosen to nest as part of a loose colony, such as favoured sites on dairy farms, it pays to be very wary of your neighbours around nesting time. On the surface, Swallows seem socially monogamous, but in reality male Swallows will, if given the chance, be promiscuous with neighbouring females. In fact so intense is their desire to father as many offspring as possible that some males are even capable of infanticide of a neighbouring brood of chicks if they think it will force the female into being receptive again. Copulating either at the nest, on a nearby roof, or on overhead wires, established pairs of Swallows will mate frequently during the small window that the female is fertile, which generally only lasts from five days before the first egg is laid until the day before the clutch is completed. Any male attempting to mate with a next-door neighbour, known as 'extra-pair copulation', will most likely only try this devious strategy once his own mate has laid her own clutch and so past her fertile period. To counter this cuckoldry, many males will guard their females during their fertile period, but in reality are unable to be present by her side the entire time. Of course any male keen to engage in extra-pair copulation will only be successful if the neighbouring female permits this behaviour to happen, and to add an interesting twist, those females that initially chose mates with a short tail will often permit a male with longer tail-streamers to mate with her. In fact most female Swallows seem to find the males with longer tail-streamers more attractive, with the result that well-endowed males are far more likely to father chicks in nests other than their own.

Although many male Swallows won't have had the opportunity to mate with their neighbours in this way, a study of chick paternity through DNA analysis found that between a third and a half of all broods will contain at least one such extra-pair chick, making it a worthwhile strategy for the long-tailed males to attempt. Females also seem to prefer males that are already paired with mates, so relatively few unmated males will have success when striving to copulate with a female that already has a partner. However, those males initially unable to secure a nest site and a mate may still have one last chance of rearing a brood if they're able to displace another male. Any unmated male able to drive an incumbent male away will then be free to destroy the vanquished male's clutch or kill his chicks. Desperate to rear a clutch and ensure her journey all the way from South Africa was worthwhile, the resident female will then have little choice but to mate with the killer and help rear their resultant offspring. To add an extra layer of intrigue into the unprincipled nesting behaviour of the Swallow, some females engage in a process not dissimilar to the Cuckoo, called brood parasitism or 'egg dumping'. These parasitic females will lay their own eggs in the nests of close neighbours also at the egg-laying stage, and providing the egg hasn't been placed into a still empty nest (which will cause it to be rejected), the unsuspecting host female will just assume the introduced egg to be one of her own and rear it alongside her own clutch.

Having fed well on the sugar-rich aquatic plants along the Russian Federation's northern coast, the Bewick's Swans will leave the sanctuary of the White Sea around the middle of May to fly north-east into the incredibly remote Nenets Autonomous Okrug (or Region). Another of Russia's federal subjects, despite this vast region being rich in oil and gas, its remote location means these resources are exploited by a population of little more than 40,000, around half of which reside in the provincial capital of Naryan-Mar. Playing host in the summer to a large proportion of the entire Bewick's Swan breeding population from the European flyway, this region is mostly composed of barren Arctic tundra, with the and adjacent to the Barents Sea being the main area favoured by the swans. As the Bewick's swans finally reach their destination, their arrival will mean that the breeding season can at last swing into action.

By mid-May, most Waxwings should have arrived at their breeding grounds in the taiga forests. Breeding in such a remote location has resulted in few studies of Waxwings in the field, so most of the behavioural observations have come from captive birds. It's presumed that those males already in established pairs will have already initiated courtship prior to their arrival at the breeding sites in order to hit the ground running. The courtship behaviour could never be described as extravagant, and is believed to consist of little more than the male forming a hump posture with his body and rump feathers ruffled, while depressing his tail, raising his crest and hopping around in front of his mate. A responsive female will then also ruffle her feathers and quiver her wings in return before allowing him to pass her a food item. This relationship-bonding behaviour is believed to continue right the way through nest building to the point when egg laying begins, and has certainly been observed taking place in northern Scandinavia as late as the middle of June.

For those birds not already paired up, it's assumed that the largely sociable lifestyle of the Waxwing means that males and females should find mates with relative ease, rendering the prolonged and noisy courtship of many other bird species redundant. This is a smart move in a location such as the taiga, where the breeding season will always be far more truncated than at temperate latitudes like Britain. Certainly in the closely related, and more intensively studied, Cedar Waxwing from North America, it has been revealed that females select partners according to the number and state of the red waxy tips on their secondary feathers, which ringers now know to be an indicator of age and experience. Not all the Waxwings travelling north will breed, of course, and nomadic flocks comprising a mix of first-year and unpaired birds, can also be seen on the breeding grounds. These inexperienced and unlucky birds will use their time to learn both the lie of the land, and the tricks which should enable them to have more success on their return the following year.

Back in Britain, the speckled Robin juveniles that managed to negotiate their first traumatic week out of the nest will still need help finding enough food for another fortnight before they can consider themselves truly independent of their parents. With spring now well and truly under way, food should certainly be far easier to find, as the youngsters learn to take advantage of the exponential increase in invertebrates that the longer and warmer days will bring.

When foraging, the Robin's favoured technique is to take food from the ground while searching through the soil and leaf litter, but they will also use any lower branches of trees or bushes as observation perches, before flying down when any movement from a tasty morsel is detected. Their search image will encompass a wide and varied diet of caterpillars, beetles, flies, ants, spiders, centipedes, earthworms, slugs and snails, all of which will be bolstered by seeds and fruit as invertebrate abundance declines later in the year. As the chicks steadily become more independent, their mother should by now already be focused on clutch number two, and the completion of her second rest will see her wasting no time in proceeding to lay her second batch.

Those Lapwing chicks managing to avoid the jaws of a Stoat, the talons of a Buzzard or the beak of a Carrion Crow will see their growth quickly accelerate as they approach the grand old age of a couple of weeks. Upon hatching, their wings were little more than articulated stumps, but flight feathers will now be emerging through the down as the young go about their feeding business under the watchful eyes of their parents. Lapwing young hatch with relatively large eyes and a well-developed head, bill and legs, meaning they are well equipped to both spot and quickly dispatch a whole variety of beetles, fly larvae, earthworms, caterpillars and spiders. However, it won't be until early June that they will be ready to fledge, and a further week after that until they would be able to stand any chance of survival without their parents' constant support.

Despite flying well just two weeks after leaving their nesthole, the Tawny Owl youngsters, unlike the more precocious Lapwing chicks, will still be entirely dependent on their parents for all their dietary needs right the way though to at least midsummer. How many chicks successfully fledge will depend on a variety of factors: the quality of the territory, their parents' level of experience and perhaps most importantly whether it happens to be a good or bad vole year. An extended study in Scotland found that in years with plenty of prey abundance an average Tawny Owl nest would fledge 2.6 young, in contrast to only 0.2 young per nest in years when voles were thin on the ground. For those Tawny Owl pairs nesting in urban environments, and more heavily reliant on bird prey, there is thought to be less of a cyclical element to the number of young successfully fledging each year, as few bird species exhibit such pronounced fluctuations in their population levels when compared to small mammals. Irrespective of how many chicks successfully fledge, those young that have already beaten the odds by surviving well into May will at this stage be relentlessly hectoring their parents for food, as mum and dad work hard to ensure their young are sufficiently well fed.

By the time the Peregrine chicks are around three weeks old, they will rarely be brooded by their mother unless the weather suddenly takes a severe turn for the worse. They will also have grown considerably in the short space of time since hatching, and even at this relatively early stage, the larger female chicks can often be differentiated from the smaller, lighter males. Their contour feathers (the outermost feathers providing the colour and shape of any bird), including those belonging to the wings and tail, will also be growing rapidly and should by now be visible as they emerge from the chicks' second coat of down. Starting off looking just like light-blue pins, as the feathers emerge from the sheaths they will uncurl and straighten, with the nestlings' heads often showing the characteristic juvenile markings quite early in thier development.

When not sleeping, the chicks will be far more active, with wing-flapping, stretching and preening all part of their daily routine as they begin preparations for life away from the ledge. At around four weeks old, the chicks will also have found their voice, and can be heard either alarm calling with their parents, or making a racket particularly when food arrives. By the time the youngsters reach a month old the female should be back actively hunting again in an attempt to keep up with their insatiable demands, and the arrival of either adult is often the cue for pandemonium to break out as the ravenous nestlings crowd around in an attempt to grab any prey brought in. Whichever chick is successful in snatching the spoils will then often use its wings to 'mantle' or hood the item, primeval behaviour that has evolved to enable the youngster to eat without interruption. Sometimes a tug-of-war over a prey item can ensue, but once possession has been decided, the victorious nestling is usually left to eat its ill-gotten gains in peace. The number of feeds can vary between around six to eight a day during this demanding period, with food brought in any time from before dawn to after dusk. During this time of plenty, the male may also begin to cache food whenever there is a surplus, as he prepares for the leaner times that will surely lie ahead.

Growing quickly on a constant stream of fish, the Kingfishers' nesting chamber by now will not just reek of partly decomposed and regurgitated meals but also look a sight, as the discarded sheaths are added to the overall mess. The youngsters, when not cadging fish from their parents, will spend a lot of time preening their newly acquired feathers, and where room permits, may also be stretching and flapping their wings to build up the all-important flight muscles. As fledging approaches, the nest can often become a much quieter place as the parents slowly reduce the frequency of feeds in order to encourage the chicks out. As the middle of May slips by, the bright orange and blue youngsters will intuitively and suddenly know it's time to leave as they trudge down the tunnel and into the light for the first time.

Due to the delayed arrival of spring further north, Scottish Blue Tits may well only just be beginning incubation, as their southern cousins see their clutches starting to hatch. Immediately upon emergence of each chick the female will step into pick up any discarded eggshell, before then jettisoning it some distance away from the nest. The entire clutch can take anything from a few hours to a whole day to hatch, and as the tiny, blind chicks are susceptible to chilling at this stage they will need to be initially brooded for large parts of the day and all night. The clutch hatching is also the trigger that instantly sees the male pressed into action as he undertakes the first of many foraging trips for his brood. Initially tiny morsels like aphids, small caterpillars and spiders are brought back, but the young will quickly be in a position to start taking caterpillars, the staple diet of any Blue Tit chick. To keep up with the increasing demands of their ravenous youngsters, the number of visits to the nest by the male, and in good weather by his partner, will rise exponentially as the surrounding foliage is scoured for caterpillars. These prey items are then brought back to a sea of hungry gapes at the nest, with any faecal sacs taken away afterwards as the parents waste no time in departing on yet another foraging mission. If the adults have timed their broods correctly, they will be able to take advantage of the sudden huge flush of caterpillars that late spring brings, and with so much natural food on offer in the countryside, garden bird tables will often become surprisingly quiet when compared to winter and early spring. While sunflower hearts or peanuts will undoubtedly provide sufficient energy to help keep the parents busy from dawn to dusk, they're not considered the ideal foodstuffs for the growing chicks, which given the choice will plump for a protein-rich caterpillar every time.

Late May

Safely entombed in their fortified subterranean chamber, a high proportion of all the Kingfisher chicks that manage to hatch should successfully make it to fledging day. But surely the far sterner examination of the youngsters' ability to defy the odds of reaching parenthood will only begin once they learn how to catch fish. Like many bird species, the young all tend to fledge in one day, often with gaps of 10 to 20 minutes between each youngster's leap of faith. For those that don't manage to fly immediately to safety from the entrance hole, and tumble down into the water, there is a very real possibility that unless they can flap to the bank side they will quickly drown. Once the young do manage to make it to a riverside perch, they won't return the nest chamber again, but will instead have their valiant efforts rewarded as their parents either bring them a meal to their new perch, or whisk them off for their first fishing expedition.

If the young naive Kingfishers are to have any chance of pulling though then they must hone their fishing skills as quickly as possible, with many fledglings pressed into diving on the very day they fledge. Their parents' practised technique makes a physically demanding and technically accomplished skill look effortless, and so as the youngsters attempt to emulate the 'professionals' there will inevitably be a fair number of mishaps along the way. Some juveniles, having dived in, may simply become waterlogged and drown, while others will struggle, yet somehow live to dive another day. Of course during this period of intense tuition, the juveniles will also be fed by their parents who are able to quickly locate their young charges amongst the bank-side foliage by their '*chip, chip*' begging call. This 'safety net' will not last long though, particularly as the parents will be keen to raise at least one more brood and so from around day four the adults' response to begging will change from that of feeding to rebuking. This 'tough love', however, is merely a precursor to even rougher treatment, as the parents shortly afterwords harden their attitude even further to mercilessly chasing them out of their territory. Evicted irrespective of the state of their fishing skills, the youngsters will need a fair bit of luck and to learn quickly if they're to survive through to next spring without either starving, drowning or being predated.

Unlike young Kingfishers, juvenile Lapwing are able to feed independently almost immediately upon hatching, although the Lapwings' main issue is not a lack of food but avoiding the high levels of predation seen on many wader breeding grounds. For those Lapwing chicks that hatched in late April, while still not quite able to fly, a much enhanced mobility should already have exponentially increased their chances of successfully fledging next month. The last feathers to emerge will be those from the tail, and the paler tips to the greenish-bronze feathers across their back and wings will give the youngsters a distinctly 'scalloped' appearance as they continue to feed under the watchful eye of their attentive parents.

Some three weeks after leaving the nest, all the young Robins should by now have fully graduated from fledging academy, but with their trademark red breast not due to make an appearance until later in the summer, they will still look utterly different from their parents for a while yet. This decidedly speckled and camouflaged plumage will not just help them avoid detection by Sparrowhawks, Stoats and cats, but also give them a free pass under the radar of any territorial adult Robins. Now temporarily released from the responsibility of caring for his brood, the male Robin will be free to return to his mate, who should be incubating her second clutch, which if all goes according to plan, will hatch in early June.

Having already emerged from their tree-hole cavity around three weeks ago, those Tawny Owl youngsters managing to successfully navigate the difficult first couple of weeks should by now be flying much more strongly, leaving them free to investigate their parents' territory. The boundaries marking the full extent of their parents' universe will be learnt quickly as the youngsters immediately shy away from any neighbouring adults or sounds of any broods in adjacent territories. While making their first exploratory steps, the young will still have made little, if any, attempt to catch their own food, remaining utterly reliant on their parents for all their meals. Even with the prolonged level of support given by mum and dad, it's thought that their care is restricted purely to the provision of food and defence against predators, and there seems little evidence that the adults actually train the young how to hunt on their own. This is a skill, however, that the young will not be able to put off practising for much longer as once they disperse into the surrounding countryside later in the summer they will very much be on their own.

For those young Peregrines hoping to fledge in early June, late May should see the nest site as a hive of activity. As the juveniles' wing and tail feathers develop strongly, the contour feathers will also begin appearing in lines and patches around the back and breast, leading to the preening of all these newly emerged plumes turning into a major occupation. The youngsters will also be ruffling their feathers on a regular basis, to help shake out any loose down and ensure the feathers are properly aligned to ensure maximum insulation during any cold and wet spells. The young will also be rapidly increasing their exercise regime by walking around and wing-flapping as they work on strengthening their flight muscles. This increased mobility will in turn enable them to defecate over the edge of the tower block or cliff face, so helping to make sure their freshly emerged feathers don't become prematurely soiled. As the food continues to be brought in, by around their 31st day the chicks should be capable of ripping up any plucked and headless prey by themselves, and by the time fledging arrives, at around 40 days, they should be competent in demolishing an entire fresh carcass. As the young will soon discover though, actually catching their own prey will be a different ball-game altogether.

In the Blue Tit nests of southern Britain, late May should see the nest activity reach fever pitch as the chicks' insatiable appetite increases the demand on their beleaguered parents. As the chicks are less likely to need brooding by now, both mother and father will be press-ganged into working from dawn to dusk to ensure enough food is collected, with a brood of ten chicks estimated to dispose of close to 1,000 caterpillars a day! Feeding at a rate of possibly a visit every 90 seconds will soon see the adults' plumage looking pretty ragged, or effectively like they've just been dragged through a hedge backwards – which in many ways they have! The items and quantity of food brought in to feed the chicks will vary according to where the nest is sited, with brood sizes in farmland and gardens often being lower than for those pairs nesting in prime deciduous woodland habitat. For any pairs rearing chicks in or close to mature trees, the caterpillars of the Winter Moth and Green Oak Tortrix are thought to provide a substantial component of all the food brought in to the chicks. In some years the sheer number of all the caterpillars munching away in the canopy is so high that the sound of the larvae's frass (or excrement) falling to the ground can almost sound like rain. With each oak tree thought to be capable of holding as many as 100,000 caterpillars, the first flush of foliage can even be completely stripped on some occasions. Once the Winter Moth caterpillar reaches full size on a diet of fresh young leaves, it will then lower itself down on a single gossamer thread to pupate on or below ground. This contrasts with the Green oak Tortrix moth, which always pupates in a rolled-up oak leaf.

In the years when both caterpillars are abundant and the Blue Tits have timed the hatching of their broods to perfection, many of these caterpillars will not make it to the pupal stage, as their development becomes rudely interrupted by being plucked off a leaf and subsequently rammed down a hungry Blue Tit's throat! So important is this brief annual harvest for Blue Tits that in years when numbers of caterpillars are lower, or the birds' timing is a touch out, many chicks will simply starve in the nest. In those broods receiving plenty of food the young will quickly feather up, with the nest quickly beginning to look like it's about to burst at the seams and decidedly unfit for purpose. There will also be plenty of stretching and whirring of wings at this stage as the young Blue Tits dispense with the last of the down and build up their flight muscles. Knowing full well where the entrance (and exit) hole is due to constantly watching the comings and goings of their parents, after just a couple of weeks the chicks won't be able to resist jumping up for their first views of the world beyond the confines of their nestsite or nest box.

Finally, some 3,500km away from the nearest British Blue Tits ready to fledge, the Bewick's Swans will at last have begun to arrive at their remote Arctic breeding grounds. Until recently, very little was known about the Bewick's Swan's summer haunts, as the Nenets Autonomous Region has only recently opened up to western ornithologists, and even for Russian scientists, the vast, sparsely populated terrain makes visits for any length of time difficult. Those Bewick's Swans overwintering in Britain are, in essence, thought to breed in a broad strip along the Barents Sea coast, anywhere from the Malozemelskaya tundra in the west, right through the Bolshezemelskaya tundra and as far as the Yugorsky Peninsula some 950km further east. This huge expanse of

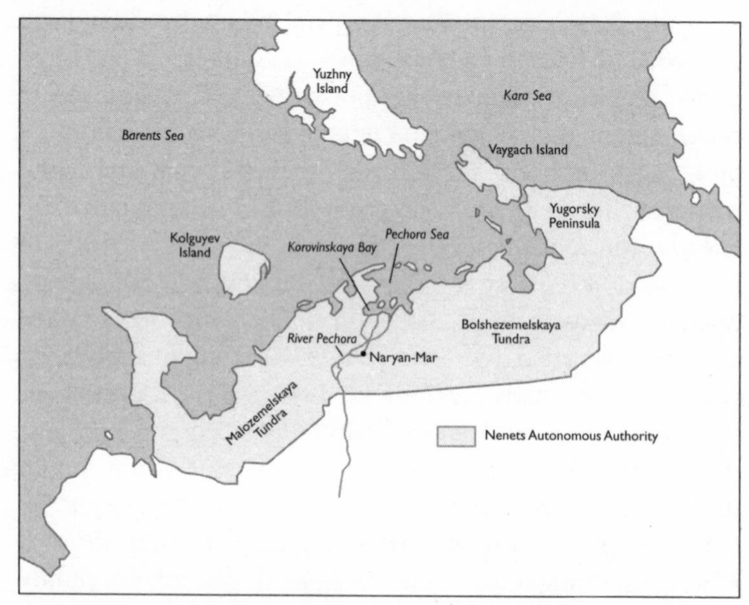

The known breeding, moulting and pre-migratory sites of Bewick's Swans in the Arctic region of the Russian Federation.

land entirely confined within the Arctic Circle mostly consists of maritime tundra and low-lying marsh that is riddled with pools and lakes, criss-crossed by river channels and dominated by a mosaic of moss, lichens and sedges.

Any offspring reared during previous breeding seasons, and which spent the winter alongside mum and dad, will by now have already dispersed giving the parents the opportunity to breed unencumbered. One crucial factor which strongly determines individual nesting success in this harsh, unforgiving climate will be the age and experience of the birds. As most pair-bonds don't even become established until at least three or four years of age, it's distinctly possible that many pairs will not even attempt to breed for the first time until even older than this. Researcher Eileen Rees from the WWT has made a study of Bewick's Swans on their summering grounds and believes only between 20% and 64% of established pairs will nest each season, with the climate ultimately dictating collective breeding success. In those years when spring arrives later and the summer ends up being colder than average, clutch sizes will additionally tend to be smaller. In fact, upon arrival, a number of pairs may decide an attempt is not even worthwhile and will not even bother holding a territory, but will instead join up with those birds still too young and inexperienced to form gregarious non-breeding flocks on shallow lakes and marshland nearby.

However, those pairs keen for their journey not to have been wasted will suddenly begin actively shunning the company of other swans, as the most experienced pairs return to locations that led to successful breeding attempts in previous years. Invariably they will arrive to find their old territories still covered in snow and ice, but these conditions won't stop them marking out the boundaries and actively defending their reacquired real estate from any swans they perceive to be either trespassing on the ground or using their airspace. Mating is thought to occur close to, or on the breeding grounds, and the displays prior to copulation include subtle ritualised head-dipping, followed by the birds rising out of the water with their wings extended, while calling vigorously. With so many predators roaming the tundra, their next job of deciding the specific location for the nest will be crucial, as the brief Arctic summer will not afford many second chances to those pairs with either chilled or predated clutches.

Like the Bewick's Swans, the sparsely populated areas where Waxwings nest means the breeding ecology of this species is also little known. In locations with plenty of suitable habitat, there seems to be a colonial element to their breeding, with the distinctive sociability that is normally such a feature of this bird in its winter quarters, not breaking down completely when nesting commences. In studies of captive Waxwings, the male was observed to select the nest location and the few observations of wild birds made suggests they favour old, stunted and lichen-festooned conifers, choosing anywhere from three to 15 metres above the ground to rear their young. As many of the birds will have already paired up before reaching the breeding grounds, the male will waste little time in declaring his territory from a tree-top close to the nest with his simple high-pitched trill. This statement of intent may also serve to encourage the female to cement their bond by joining him for a duet.

Mating is thought to occur both close to and on the breeding grounds, with the male continuing to feed his mate a combination of fruit and insects to ensure she will be in good enough condition to produce her clutch. Built by both the sexes, the female is believed to assume the larger role in nest construction as the base is built up with a cup of thin twigs, before being lined with Reindeer Lichen, moss and fine grasses. The period between when nest-building is finished and incubation is initiated is thought to be the only time when the normally placid males show aggression towards their neighbouring rivals. During this window any intruder perceived to be interrupting business at will be sent packing by a combination of bill-gaping or snapping, as the incumbent male presses home his advantage without having to resort to physical violence.

For any Cuckoo chick that has successfully evaded detection in its foster parents' nest throughout incubation, violence will quickly become order of the day upon hatching. Weighing in at little more than 3g, the Cuckoo's egg has been carefully sculpted by evolution to mimic both the size and pattern of the chosen host species' eggs as closely as possible. Incubated in amongst the host bird's clutch, the alien chick will also be blessed with a head-start over the rest of its host's brood. Having been laid with a partially developed embryo, means the Cuckoo chick could hatch in a remarkably quick 12 days after its mother's clandestine deposition. Hatching earlier brings considerable advantages to the Cuckoo chick, as the host's own clutch of eggs, or very young chicks, are certainly much easier to eject from the nest than, say, older host chicks, which may have already put on a substantial amount of weight.

The credit for the first detailed description of Cuckoo chicks ejecting host eggs, or young, remarkably goes to Edward Jenner, the celebrated scientist who effectively invented vaccination. Before his seminal work on smallpox, he published a paper on the Cuckoo's nesting habits in 1788, which led directly to his election to the prestigious Royal Society the following year. At the time his assertions were met with widespread disbelief and initially rejected; however, as other naturalists subsequently backed up Jenner's observations he was quickly vindicated. What Jenner first recorded must surely go down as one of the greatest Machiavellian feats in the natural world.

It's generally not until around eight or ten hours after hatching that the Cuckoo chick will have garnered sufficient strength to begin the devious business of ejecting the host eggs or chicks. Despite being naked, pink and blind the chick has a broad back complete with a shallow depression, and by wriggling around in the bottom of the nest at some point one of the host's eggs or helpless chicks will finally come to rest on its back. By then using its legs as a brace against the side of the nest, and its wings to hold the egg or host chick in place, the young Cuckoo will then begin the exhausting process of hauling its cargo up the wall of the nest. Working in short bursts with frequent rests, as soon as the egg or chick finally reaches the rim, one last quick jerk will see it pushed out of the nest, and in the case of the Reed Warbler, straight into the water below. In the nests of small hosts such as Reed Warblers and Meadow Pipits, it might take less than 20 seconds to evict an egg, but having seen this happen many times, Cuckoo expert Nick Davies reckons the average time taken is around three to four minutes per egg. Needing also to recoup its energy in between evictions, the disposal of an entire clutch might take anywhere from three or four hours to a couple of days if the chick has to eject large host young. After four days, this 'ejecting behaviour' then suddenly disappears, by which time the Cuckoo should have become the last chick standing! Oblivious to this blatant act of skulduggery, the unquestioning host parents will then simply devote their entire attention, and food supply, to the Cuckoo in the nest.

Hatching out in southern England, Nightingale nestlings certainly don't have the killer instinct of a Cuckoo chick, and compete on a fair and level playing field with their fellow siblings for the food brought in. Instinctively begging soon after hatching, the young are initially brooded by their mother, leaving the male as the sole provider of food during the early stages. Like Blue Tits, the Nightingale pair will aim to synchronise their brood to the precise moment when concentrations of caterpillars are at their highest. Any tiny caterpillars or other morsels brought in by the male are passed initially to his mate, who will in turn feed the small, naked and blind young. This arrangement won't last long, and as the chicks' appetite grows, both parents will be forced into foraging for larger caterpillas to keep up with the demand.

The British naturalist Oliver Pike, in his book *The Nightingale: its Story and Song*, published in 1932, spent many hours watching Nightingales on their breeding grounds. He reported on one nest where the adults came in every minute for nearly three hours and the four young were always ready for their meal. Before the young's eyes were open, Pike observed the female making a low '*zee-zee-zee*' call to prepare the young for her arrival, but later on as the chicks were able to see, this call was rendered redundant, as beaks suddenly shot up at the first sight of her return with food. Watching the nest from his hide with a periscope, Pike didn't think that the parents travelled far from the nest and that all the food was collected strictly from within their territory. Although the male will still sing at this stage, it will be only sporadically, with any bursts heard presumably giving male chicks in the nest the opportunity to listen and start learning the complex songs they will need to have mastered by the following year – providing they manage to both fledge and successfully negotiate a return trip to Africa!

By now in full laying mode, female Swallows will be aiming to produce an egg each morning until their clutch is complete. The number of eggs can vary between two and eight, but a clutch of four or five is by far the most common. As the eggs are formed directly from the female's current intake of food rather than her fat reserves, the prevailing environmental conditions are thought to impact on clutch size. Each weighing in at around 2g, the eggs are usually white with varying amounts of reddish- or purplish-brown speckles and blotches. Only starting her incubation with the penultimate egg, as the female alone has a brood patch, on the few occasions where she disappears off to feed the male won't normally sit on the eggs. Angela Turner reported that an incubating female will generally spend 60 to 80% of the day incubating her clutch, and as the male doesn't feed her at the nest, she will need to intersperse these sitting duties with short feeding bouts lasting anywhere between two and nine minutes. Her incubation is of course carefully timed to coincide with the first large proliferation of insects in spring, meaning she shouldn't need to be away from the eggs for any longer than is absolutely necessary. On particularly cold days, however, the female will need to spend more time hunkered down with the eggs, and when eventually forced off by hunger, she will just have to hope that the nest's feather lining will help ensure the clutch doesn't cool down too much before her return.

Having laid their single precious egg in early May, come the end of the month the Puffins will still have some way to go before their chick or 'puffling' will be finally ready to make its entrance into the underground world. Despite needing around six weeks for incubation, which is a remarkable length of time for such a small bird, the relatively constant temperature underground should give the pair enough time for short socialising breaks. When the pair are both present at the puffinry early or late in the day, they will rarely miss the opportunity to indulge in a spot of ritualised behaviour, such as 'billing', the 'pelican walk' or 'spot-stomp'. Practised by generations of Puffins, these actions all reinforce the same message to any neighbouring birds looking on: 'This is my mate for life, we have a strong bond and this is our burrow, so please keep your distance!'

June

The middle of the year brings a wonderful blend of the luxuriance of spring and the warmth of summer to the British Isles. As the sunshine warms the soil this is the time for huge floral diversity and is without doubt the best month to look for the most enigmatic and beautiful of all the plant groups – orchids. With flowers resembling monkeys, ladies and lizards, it's not just the orchids that will be the centre of attention for insects, as most colourful blooms may well have a whole swarm of pollinators in attendance, keen to take advantage of the nectar and pollen on offer. With the summer solstice also happening this month, the long days will provide ample opportunities for the many birds now fledging to find sufficient food while learning the ropes away from the confines of the nest. For some youngsters, the extended parental care can last all summer, but for others it will be a steep learning curve as they are given the cold shoulder by mum and dad, keen to turn their attentions to brood number two.

Early June

After around 20 days cramped up in the nest, in quite possibly stifling conditions while competing for every caterpillar that arrives, many juvenile Blue Tits will decide they can stand the conditions no longer and begin to fledge. With most Scottish Blue Tits only just beginning to hatch at this point, it will definitely be the southerners leading the way, as the first chick jumps up to the nest entrance and grabs the rim of the hole with its feet before squeezing through into a world full of infinite possibility and danger. Fledging usually takes place early in the morning to give the birds all day to become accustomed to their new surroundings, and as the first bird leaves, this often gives impetus to the other chicks to seize the moment and cross the line from nestlings to fledglings.

Their parents will invariably be close by during this defining moment of the breeding season, and play their part with encouraging calls as the chicks are enticed out and hopefully up into the relative safety of the trees. With their distinctive bluish-green caps, yellow faces and still a hint of a yellow gape, juvenile Blue Tits immediately stand out even to the novice birdwatcher. Unsurprisingly, given the restricted room in the nest, their wings are still not fully grown, so upon fledging their flights are initially short, feeble and clumsy. Despite the best efforts of the parents to keep their young fed during this period, the harsh reality is that as many as 90% will not even see their first birthday, with the first few days out of the nest exacting the heaviest toll. Any young Tit grounded will immediately be at great risk from predators such as Weasels, Stoats and of course domestic cats. The youngsters will also have to remain alert above ground too, as Sparrowhawks will make short work of a defenceless Blue Tit with their own chicks hungrily waiting back at the nest. The other big factor affecting the survival of young and inexperienced birds like Blue Tit chicks will be the weather, and any prolonged cold and wet periods immediately after fledging can have a dramatic effect on the number of juveniles able to see the summer out.

Close to six weeks after hatching, and some 78 days since the first egg was laid, early June should see Peregrine chicks lifting off, as all their practice flapping enables them finally to defy gravity for the first time. Due to the elevated temperatures in towns and cities caused by the heat island effect, urban-nesting Peregrines may fledge a week or so earlier than their countryside cousins. As some of the chicks' feathers are still in pin, they will often be a touch overweight on leaving the nest, meaning that their first flights can be clumsy and haphazard. Certainly any ecclesiastically-bred birds that don't manage to flutter to a nearby spire or gargoyle on their maiden flight could easily become grounded in the graveyard below or even further afield. Often the smaller and lighter males tend to fledge marginally before the bulkier females, and those lucky enough to have grasped a suitable perch close to the nest will then be content to stay there for a few hours while noisily haranguing their parents into rewarding their efforts with something edible. Within the space of just a few days the young should have both strengthened their flight muscles sufficiently and lost enough weight to make flying much more straightforward, but it will be far longer before they will manage to master the dark arts of catching moving prey!

Also taking to the air for the first time, any surviving Lapwing chicks able to simply airlift themselves away from danger will have suddenly and massively boosted their chances of making it to adulthood. It has been calculated that Lapwings only need to rear between 0.8 and 1 chick each season to sustain their numbers, and the very fact that the British breeding population has fallen by over 50% in the last 25 years shows how a combination of agricultural intensification and increased predation levels are severely impacting the survival levels of this wonderful wader. Even after the young have begun flying they will remain dependent on their parents for at least another week, and can still be clearly picked out by their shorter crest, less well-marked face, scaly backs and incomplete breast-band. By the time their youngsters have begun to fledge, the adults will have already started their annual moult. But needing to remain airborne, in order to track down widely dispersed feeding opportunities throughout in the summer, it is likely the moulting process will not be completed until later in the autumn.

In the nest for a grand total of no more than 13 days, the Nightingale chicks will need to grow quickly on their non-stop diet of invertebrate food. As fledging approaches the feathered youngsters will suddenly become much more aware of their surroundings and crouch down if danger threatens. As a last line of defence they will have also developed the ability to gape in a threatening manner if any predator stumbles across the nest. The parents will be only too aware that the success of their entire breeding season will be riding on this one precious brood of chicks and so will give excited alarm calls and mob any potential predator zeroing in on the location of the nest. As a last resort the adults will also attempt a distraction lure, where they will try to distract the predator by feigning injury, such as a broken wing, while fluttering around on the ground in a seemingly hopeless fashion. The predator – thinking an easy meal has presented itself – will be drawn towards the adult and away from the chicks, only then to see the adult Nightingale make a miraculous recovery at the last second.

Leaving the nest, the spotty and brown young Nightingales look superficially similar to fledgling Robins, and being still unable to fly for a further three or four days, will need to be quickly led away by their parents into the comparative safety of the surrounding undergrowth. Unlike the majority of their continental cousins, British Nightingales are only thought to have one brood, so with both the adults freed from any further breeding constraints they're able to devote themselves to looking after their young for at least the next fortnight. Still reliant on their parents to find them food for at least the first week out of the nest, it's thought that the brood may often be separated between parents, giving each youngster a better level of attention and more personal tuition in the ways of finding food. Dividing the chicks will also split the risk, thereby maximising the chance that at least some fledglings will make it through to autumn migration.

His duties suitably discharged when his first brood disperses into the surrounding countryside, the male Robin will waste little time in turning his attention back to his mate, who should by now already be sitting on clutch number two. As the male doesn't generally feed the female while she incubates, this will give him a little time to catch his breath before the second brood begins demanding food. He will of course not abandon his responsibilities entirely during this down-time, and be on hand to safely whistle her off the eggs if and when she needs to feed. If his mate is tucked away on a nest with minimal visibility of her surroundings and a mammalian predator (such as a human)

is observed too close, the male will often make a '*tic-tic*' alarm to warn her, which once learnt is easily recognised. Birds such as Sparrowhawks are treated differently, as these represent not just a very real threat to the female herself, but also to the entire clutch if she were to be predated. If a Sparrowhawk is spotted close by then the male will make a '*see-eep*' call, warning her to keep her head down and stay still. If all goes according to plan, and after only two weeks hunkered down on the nest and a mere seven weeks since her first brood hatched, the Robin's second batch of chicks should then begin to emerge into the nest. Upon hatching they won't immediately need feeding while they use up the last of their yolk reserves, but the removal of the broken eggshells from the nest will surely serve as a clear statement to the male that his relaxing break has just come to a shuddering halt.

Still with their first brood, the river can be a noisy place as the Kingfisher parents mercilessly force their young out of the family's territory. Armed with little more than just a few days spent watching their parents catch fish with unerring accuracy and effortless skill, the evicted youngsters will have little choice other than to quickly spread out along the artery of waterways surrounding what was temporarily their home. With no knowledge as to where each watercourse leads, or which territories are already occupied, these can be incredibly testing times for the young naive birds. Usually within the first day of being chased away the young birds could be at least 300 metres away, and may have moved as far as a few kilometres from their parents' territory in the space of just a few days. Those sites with plenty of fish and many potential nesting locations may see territories very close to one another, and so any youngster fishing on another pair's patch will need to keep very quiet if it's to avoid being spotted and subsequently driven out. However, if the young bird's presence goes unchallenged, it may have landed in a vacant lot, and so proceed to staking an ownership claim. Any territories secured early by young Kingfishers will enhance their survival prospects no end, as catching sufficient food will only become more difficult as summer fades into autumn. The stark reality is that as few as 20% of all fledging Kingfishers will survive long enough to see their first birthday. But if they do survive this first tough year then the odds will start to improve, although not by much, as the BTO estimates that the typical lifespan of a Kingfisher is little more than a paltry two years. In fact the oldest Kingfisher recorded in Britain was four years, six months and 13 days when re-trapped, and while older birds undoubtedly exist, the Kingfisher is certainly in the 'live fast, die young' category.

Being such short-lived birds means it certainly pays to try and rear as many young each season as possible, which in the case of the Kingfishers can be achieved by attempting a second, and even very occasionally, a third brood. Those pairs successfully managing to fledge their first brood according to plan will be keen to nest again with all possible haste. Faced with this scenario, their first decision will then be whether to reuse the previous tunnel or attempt the time-costly, but more sanitary option of digging a fresh tunnel and chamber. If the pair decides the first tunnel is still fit for purpose, then the female will obviously need to wait until the first brood has fledged before giving the nest a spring clean. But if the adults do decide a fresh start might be better after the fledging of their first brood, then some time can be saved if the first brood's diving lessons are delegated to her mate. Additionally, the second nest invariably seems to be excavated more quickly, possibly as the pair will by now have gained a good understanding of how to work with the particular soil type, which should enable them to dig more effectively.

Unlike Robins and Kingfishers, the long incubation period and extended adolescence of the Tawny Owl youngsters mean that their parents will never be able to rear more than a single brood each year. Despite a month having passed since the young first ventured out of the nest and into the trees, they will still be haranguing their parents to bring them food each night, behaviour that will in all likelihood continue right the way through the summer. With their breeding season now over, and in spite of the extra demands placed on them by their needy offspring, the adult Tawny Owls will by now have begun their annual moult. Keen to keep flying and yet not compromise their hunting ability throughout this energetically costly process, the shedding and replacement of particularly the flight feathers will be carried out in a slow and methodical fashion, possibly not being completed until early autumn.

Having commandeered the nest for itself, the Cuckoo chick's next job will need to combine persuasion and trickery if it is to dupe the often much smaller hosts into the demanding business of keeping it sufficiently well fed. Naked and blind initially, the Cuckoo nestling develops a very rapid call, which is thought to fool the foster parents into believing their nest is full of hungry host chicks all calling at once. However, when they both arrive with food, there will only ever be one huge gape waiting to be filled. The hosts will of course bring in the same type of food they would normally feed their own chicks, so in the case of Reed Warbler hosts, for example, the diet will consist of either beakfuls of small flies or larger single items such as caterpillars, moths, butterflies, damselflies and hoverflies. As the Cuckoo chick has such catholic preferences, all offers will be gratefully received!

Eating the amount of food that would have kept an entire brood of either four Reed Warbler or five Meadow Pipit chicks more than happy, the single Cuckoo chick will grow quickly as both host parents run themselves ragged trying to keep up with the incessant demand. Although pink when hatching, by the time the youngster is three days old it will have turned black all over and developed a vivid orange-red gape to spur its foster parents on with their misguided mission. On the fourth and fifth days the feather quills begin to emerge, giving the youngster a dark and prickly appearance, and by the time its eyes open after around a week, the chick looks nothing whatsoever like the host chicks would have looked were their lives not cut so brutally short. By the time the feathers begin to emerge, the chick will have grown so quickly that the foster parents will be unable to brood their charge. In addition to being quite bulky, the chick will also be able to look after itself and can adopt an amazing defence if threatened. Erecting its feathers and opening its bill to reveal a huge bright red gape, it will then suddenly stretch and snap its neck back in a surprising, intimidating and aggressive manner. Any potential threat promptly dealt with, the Cuckoo chick will then divert its attention straight back to the matter in hand – cajoling its foster parents to hurry up with more food!

As the Cuckoo chick continues to throw its weight around, many Swallows will be busily preparing for the moment their first broods start hatching. Emerging at around 18 days, all the chicks will hatch within a 24 to 36 hour period, and the first act of the female will be to make some more room for the youngsters by tossing the empty eggshells just a few metres away. At hatching, the Swallow chicks will generally weigh less than 2g and apart from a few wisps of grey down on the head, back and wings, emerge naked and blind. In common with many other chicks, for the first few days they're unable to regulate their own temperature and so must be brooded by the female almost continually. During these early days the chicks' food requirement will also be minimal, so the male should be more than able to cope as he brings in the first supplies.

At this stage the chicks are thought to only know when food is imminent by their parents' contact calls and possibly by the shadow cast as the male approaches the rim of the nest with food. Even at a very tender age, the Swallow chicks' gape is a prominent red-orange in colour and is surrounded by a paler yellow flange. Many birds see in ultraviolet light and as the lighter border reflects a different part of the spectrum to the gape itself, this helps guide their parents into the 'feeding zone' in the dim light, much like the landing lights on an airport runway at night.

Having laid well before the vast majority of Swallows, the Puffins' single egg will only now be finally entering the last stages of its six week incubation. Taking three or four days to chip its way out of the egg there will almost certainly be an element of communication between parents and chick as it attempts to break out, with the message from inside the egg being 'Get ready to start feeding me' coming through loud and clear!

Bringing up the rear in the mating stakes, early June should finally see summer belatedly arriving for those intending to breed at higher latitudes. Exactly when Waxwings and Bewick's Swans begin to lay their clutches will depend on how cold the previous winter was and the exact timing of spring's arrival in the frozen north. In an average year early June should see the Waxwings putting the final touches to their nest, while the male continues to feed his mate to ensure she will be fit and able to lay. Due to their much longer incubation time, the Bewick's Swans can't afford to waste a moment, and so those pairs that have successfully secured a territory should already be actively laying by now. The exact position of the swan's nest is considered crucial, and with no option of rearing their brood anywhere other than on the ground in this barren landscape, an experienced pair will, wherever possible, choose a raised, south-facing hummock or mound. Slightly elevated nests tend to serve the swans best as they give a good all-round view, enabling the pair to track the movements of predators across the tundra, and those locations with a south-facing aspect will also become ice-free quicker.

Once the spot is chosen, the mound is built by simply piling up vegetation from the immediate surroundings, so by the time they're finished the nest may extend over a metre across and reach a height of up to 50cm. Looking like a doughnut, the central area – which will hold the precious clutch – is then lined with a mix of softer grasses and any down the birds might have going spare. With a good territory so vital for breeding success, many experienced birds will not only return to occupy the same location, but even reuse the exact same nest from previous seasons, after any necessary running repairs have been carried out, that is. Immediately after laying, the eggs are white in colour, but then stain to a brownish-yellow after a few days, and while not quite up to the size of the huge eggs laid by Mute Swans, each Bewick's egg will still top the scales at around an impressive 280g. Clutch size will not only vary across the tundra according to the age and experience of the pair but also from year to year, however nests of between three and five eggs seem most common. Once laid, incubation will begin immediately and with both parents doing their utmost to protect the clutch from both predators and the worst the weather can throw at them, they will hope to see their cygnets for the first time in just under a month.

Mid-June

The sight of Puffins bringing fish ashore will surely mean just one thing – the chicks, or 'pufflings', have begun to hatch. Finally, after an incubation of around 41 days, which will have included three or four days of hard labour as the chick breaks free, the bedraggled puffling will emerge into the dark underground world that will represent its entire universe until it heads out to sea later in the summer. Another sign that pufflings have begun to hatch left, right and centre will be the sudden accumulation of remnant eggshells at the various burrow entrances, signifying completion of the first step in a life that, if all goes according to plan, could last over 30 years.

After drying off, the chick will resemble a charcoal-coloured powder puff, with lilac-coloured feet and a black stubby bill tipped with the white egg tooth, which might stay attached for a couple of weeks. For the first week after hatching, the chick is unable to maintain its body temperature and so would quickly chill and die without being brooded continuously by one of its parents. Rarely needing to eat on hatching day as it finishes off the last of the egg nutriment inside its body, it will quickly develop a keen appetite as the first food items are transferred carefully from the parent's bill to the chick's ready gape. Like most chicks on a rapid weight gain course, the puffling will initially spend large parts of the day sleeping, only becoming more animated once a little older.

The young Puffin's diet consists of pretty much the same food as their parents, with typical prey being a combination of Sandeels, Sprats, small Herring and young stages of fish (gadoids) in the Cod family, such as Whiting and Haddock. Of all the prey brought in, by far the most frequently caught and important item will be Sandeel, with most Puffins in British waters catching the Lesser Sandeel. These small and silvery-coloured eel-like fish have long fins, a forked tail, protruding lower jaws, and are very familiar to anyone who has watched adult Puffins returning with bills full of them during the height of the breeding season. The timing of their life cycle can vary around Britain, but certainly populations surrounding the Shetland Isles tend to spawn in December and January. The eggs then stick to sand grains on the seabed until hatching, when the larvae become pelagic and float around with the zooplankton, until ultimately transforming into the small recognisable Sandeels by May or June. As their name suggests, Sandeels spend much of their life buried in the sand, but in the summer occur as large shoals in the water column when feeding on plankton during the day, before returning at dusk to spend the night tucked up in the sandy seabed. Usually found in waters at depths between a few metres and 70m, this annual harvest is easily within diving range of any foraging Puffin with a hungry chick to feed back on dry land.

Another chick pushing its 'parents' to the limits by mid-June will be the Cuckoo. By now much larger than its foster parents, this huge size difference doesn't seem to prevent the chick's industrious hosts from realising anything is wrong as they continue to stuff the youngster with food from dawn to dusk. Quite often in order to ensure the food reaches its required destination, the host adults may even need to stand on the Cuckoo chick's back in order to reach its gape, and the nests of Reed Warblers and Meadow Pipits soon look manifestly unfit for purpose as the chick grows out of house and adopted home. Feathering up quickly, the Cuckoo chick will by now be quickly developing the characteristic white feathers on the back of its head and on spotting its 'parents' approaching with food, the distinctive hissing call can be heard from quite some distance. Upon arrival of the food, the bird seals the deal with a demonstrative begging gape and quivering of wings.

Despite eating enough for four or five, the young Cuckoo will actually take longer to fledge than the chicks of most of the host species it parasitises. Certainly in Britain, Reed Warbler and Meadow Pipit chicks will leave the nest after around 12 to 14 days and be dependent on their parents for only a further 12 days afterwards. This contrasts with the Cuckoo needing anywhere between 17 and 20 days in the nest, and still reliant on its hosts for almost as long again, before its departure finally relieves the burden from its beleaguered foster parents. After leaving the nest, the young Cuckoo will usually seek the safety of nearby bushes, where it will then continue to noisily beg for food. Looking rather like an adult Cuckoo at this stage, the fledgling will usually tend to be a touch more rufous in colour and with heavy chestnut barring on its upperparts. It also pays for the greedy youngster at this stage to remain initially largely hidden and perched in one place, as its silhouette in flight will frequently illicit aggressive responses from many local birds and even leading to it being mobbed by its own foster parents! Often extending its shameless begging repertoire to any passing bird, it seems the parental urge of other species does occasionally extend to ensuring this bare-faced cheat will continue to be fed by all and sundry until full independence finally sees it foraging for its own food.

The extended adolescence of Tawny Owls means any fledged young by mid-June will not only still be hanging around their parents' territory, but also utterly reliant on any food brought to them for some time yet. Tawny Owl expert Dave Culley has followed his neighbourhood pair through a number of breeding seasons. Breeding in a small yet wooded suburban location in Cheshire, Dave has found that in April the owls' diet consists of around 40 to 50% birds. But by June as much as 90% of all the prey caught and brought to the youngsters will be feathered. With the constant pressure to find sufficient food to feed their young, the adults at this time of year will capitalise on the easy prey of recently fledged birds, even hunting during the day when the opportunity arises. With numerous cameras set up throughout his owls' territory, Dave has even managed to film one of the Tawny parents snatching a female blackbird incubating a clutch right off her nest!

Like their Tawny counterparts, young fledged Peregrines will also be some way off being able to feed themselves by mid-June. Gaining confidence in the air will be their first task and as they slowly master the necessary skills, their first flying missions will see them relentlessly chasing their fellow siblings and parents. In those locations where three or more young have successfully managed to fledge, this can be an exciting time to watch Peregrines, as all the juveniles tear after any adult returning with food. The difference in flying ability between adults and young will initially be striking, but by steadily learning the vast array of tricks needed to survive, they will hopefully soon be up to speed. Able by now to dismember and feed on a complete carcass on their own, the youngsters will still be very much home birds and not dare to venture too far away from the site from which they hatched in late April.

Despite the oldest Robin ever recorded by the BTO being a bird from Fylde, Lancashire, which lived for over eight years and four months, average life expectancy of the nation's favourite bird is thought in reality to be a meagre two years. With such a short lifespan, it's no surprise that Robins are keen to produce as many clutches as possible, meaning by the middle of June many pairs could already be well into feeding their second brood of the year. With daylight lasting over 16 hours by this time of year, the Robins will be keen to take advantage of these long summer days to keep their brood well fed. As with the first clutch, it will take the second brood no more than 13 or 14 days before the desire to unleash themselves on the world proves overwhelming. But as the speckled and helpless young will still need help from their parents for a further three weeks after fledging, the adults will have an important decision to make – whether to stick or twist!

As many Robins see their second brood fledging, those paired-up Swallows feeding the season's first young will also be constantly on the go as they attempt to keep up with their growing chicks' insatiable appetites. The emergence of the first pin feathers at around day four or five will coincide with the youngsters becoming even more ravenous. A couple of days later on and the chicks' ability to maintain their body temperature in all but the worst weathers will then see the female press-ganged into helping her mate find enough food to support their growing brood. A study of Scottish Swallows by Angela Turner recorded the number of visits made to a brood of five chicks back in 1980. At just a day old, the chicks were fed an average of six times an hour, which was then compared with 17 visits per hour at six days old and around 29 times an hour after 10 days.

In order to pack in so many visits, most adults will hunt close to the nest location, rarely foraging more than 600m away and frequently even closer. Where Swallows have nested colonially, the feeding ranges of neighbouring birds will also overlap, particularly at traditionally rich feeding sites, such as around water bodies, or near cattle. The food the adults bring back to their chicks will also vary according to the time of year. Earlier in the season, for example, flies tend to be more abundant than aphids, and so for the first broods, horseflies and hoverflies are thought to form an important component of the chicks' diet. These large, mobile insects require more energy for the adults to hunt, so fewer can be caught on each foraging trip, but as they provide far more energy than smaller insects, it seems worth the extra effort to seek them out. When the adults are collecting insects for the chicks, they will usually catch a good number, which are packed together in a salivary ball (or bolus) in the throat, before returning to the nest. Angela Turner tried to count the number of insects in each visit, and recorded anywhere from just one insect to 126, with the average number amounting to 18. In order to rear a brood from hatching to fledging it has been calculated that the adults may need to catch as many as 150,000 insects. In addition to this high-protein invertebrate mix, the adults will also provide their chicks with grit to help grind up the food in their gizzards, so aiding digestion.

Fed on such prodigious quantities of food the chicks will grow quickly, with days three to ten being the period when weight gain accelerates the quickest. Flying in with food, it is the parents who appear to judge how hungry each chick is, with the intensity of begging a key factor, before then deciding who gets the meal. Certainly when food is plentiful, and during the course of the day, each chick is thought to receive a roughly similar amount of food. Both parents feed the chicks directly, but the male's effort can vary considerably between individual birds, with the more attractive longer-tailed males often being far more dilatory than the shorter-tailed males. This might not be entirely down to lethargy on the part of the longer-tailed males but possibly due to being less efficient foragers as a result of drag created by their long streamers.

For the first few days of the Swallow chicks' lives any waste will be removed by the adults in the form of faecal sacs, but by 12 days of age, their toiletry skills will have improved sufficiently for them to begin painting the floor below the nest. After a couple of weeks the chicks should also be well feathered, and be topping the scales at 23 or 24g, making them even heavier than their parents. After this top weight has been reached, the chicks will then usually begin to lose a few grams as excess calories are burned by the constant exercising of their wings. Needing to fly competently the instant they fledge, they will still need a third week in the nest, during which time their flight muscles must mature sufficiently for them to propel themselves into the air when the moment comes. To ensure this first flight is a success, the chicks also spend an enormous amount of time preening to make sure they will be instantly flightworthy at take off.

Unlike the busy Swallow nests, many of those which belonged to Blue Tits in southern Britain will have already been vacated as the juveniles learn the art of searching out caterpillars in the tree canopy. Over the next few weeks, with the exception of their flight and some tail feathers, much of the youngsters' hastily acquired dull plumage will be replaced as they prepare for colder times ahead. Squeezing through the hole hundreds of times to feed their chicks will also have made their parents' feathers worn and frayed, leading to many adults starting their moult even before their chicks had even fledged. But as the adults will still need to be able to fly competently for the entire duration of the moult, the whole process can last as long as 80 days and so will not be completed until the autumn. Blue Tits tend to moult in a symmetrical fashion, with rarely more than one primary flight feather missing from each wing, to ensure the birds are still able to fly in a balanced fashion. The instant an old feather is shed, a new one will then start to form from special cells in the skin. Moulting is an energy-demanding process, and the replacement of body and contour feathers will also temporarily reduce the birds' thermal insulation. So an increase in their intake of food should not only power the replacement of feathers but also keep the birds warm.

Not ready to begin their moult just yet, any adult Nightingales that managed to successfully raise a brood should currently be still helping to feed their chicks. Often staying together as a family for around two weeks at least, by the end of the first week the youngsters should slowly be starting to find their own food. Before the young fledged, the parents are believed to have foraged strictly within their clearly demarcated territory, but with their youngsters now mobile, the territoriality quickly breaks down, leaving the family free to wander further afield and to potentially richer feeding sites.

By mid-June, Lapwings should also be leaving their breeding grounds as both juveniles and adults join post-breeding flocks further afield. These flocks will provide a comfort blanket for all as the young learn the best habitats in which to feed, while the adults seek sanctuary while slowly shedding and replacing feathers worn out during the breeding season.

Breeding north of the Arctic Circle on the maritime Arctic tundra, those pairs of Bewick's Swan intent on breeding should by now be sitting on their clutch of three to five eggs. Even though the male will take turns to sit on the eggs, it seems he may well in fact be doing little more than preventing them from becoming chilled or eaten by predators, as the responsibility for most of the incubation lies with the female. Bewick's Swans are extremely shy during the breeding season, and being free from disturbance is an important factor in deciding how successful each pair will be in raising chicks. Any humans approaching to within a few hundred metres of the nest will initially cause the sitting swan to hide and if the intruder continues to move even closer this will result in the adult swan retreating to a nearby pool or lake. More natural predators on the tundra include Arctic Foxes, Wolverines, Arctic Skuas and Herring Gulls and they will often try to strike when the adults are away from the nest. Of these, Arctic Fox is thought to be the chief predator of eggs and young cygnets, and one theory believes that more Bewick's Swan eggs and young are taken in those years when the population levels of the foxes' favoured food, lemmings, are much lower.

By contrast to Bewick's Swans needing around 30 days to incubate their clutch, Waxwings will see their young hatching in just half the time. Certainly in northern Scandinavia laying normally begins in mid-June, but can occur before in those years when spring – and crucially the snow melt – occurs earlier than normal. Due to the short Arctic summer, Waxwings, like Bewick's Swans, are only ever single-brooded, and although anywhere between four and seven have been observed, the most commonly recorded clutch size is five eggs. Smooth and glossy in appearance, the eggs are pale or greyish-blue, and tend to be sparsely marked with black and grey spots that occasionally show blurred brownish edges. Like many of the smaller passerines (or perching birds) incubation is by the female alone and is only thought to begin once the clutch has been completed. Due to the fairly cold ambient temperature in June, the female will rarely leave the eggs and so will need to rely on the male to keep her nourished during her vigil with a cocktail of regurgitated fruits and insects. For the males, any brief antagonism which occurred during the nest-building and egg-laying phases should have been quickly put to one side as their natural sociability once again returns, allowing them to re-form flocks. Waxwings are generally not thought to breed in their first year, but this still doesn't stop any yearling birds returning to the breeding grounds as they join the roving flocks of breeding males. Being amongst their more learned peers will enable them to learn the ropes with the ultimate aim of finding a partner and rearing a family of their own the following year.

By mid to late June, those pairs of Kingfishers which were able to lay their first clutch as early as April may well by now be already incubating their second clutch. For the parent not sitting on eggs, its time will be divided between fishing, preening and resting. Fishing for an experienced bird during this time of plenty should be a doddle compared to the leaner times between autumn and early spring. In Britain during the breeding season Kingfishers will consume a wide variety of fish including Bullhead, Minnow, Three-spined Stickleback, Loach, Grayling, Carp, Perch and Pike – in fact any fish that have reached the ideal length of around 7 to 8cm. In the brackish waters of the Thames in London, Kingfishers have even been spotted catching small flatfish, but by far the commonest species taken are Minnow, Stickleback and Bullhead. Needing to eat roughly their own body weight every day, Kingfisher expert David Boag estimated that each adult requires around 18 fish to be sufficiently well fed. However, as soon as the second brood hatches the number needed will rise substantially!

Late June

It's astonishing to think that for those bird species rearing only one brood, the end of June will already signify that the breeding season is already drawing to a close for another year. Most traditional breeding locations of Lapwings, for example, should by now be deserted as the adults and young suddenly enter the sociable period of the calendar by coalescing into post-breeding flocks. Away from their breeding grounds Lapwing can be frequently mobile, making their movements difficult to track, but it's highly likely that any shifts in location will be dictated by the two most important factors in their lives – food availability and the weather. In addition to British-breeding Lapwings dispersing into flocks to feed and moult, the east coast of Britain may well already be seeing the first of a number of influxes of Lapwings which summered on the continent. These first immigrant Lapwings will probably be individuals that either failed to fledge young, or were simply unable to gain a territory and a mate. As the summer slowly unfolds, the numbers arriving from the Netherlands, Denmark and France will undoubtedly increase to such an extent that by the time winter has fully taken hold, the population of Lapwings in Britain will have more than doubled from around 130,000 breeding pairs to around 620,000 birds.

Commonly rearing two broods on the continent, the relatively cool summers in Britain mean Nightingales here are only ever believed to practise a 'one clutch' strategy. Having fledged earlier in the month, the families should by now be breaking up as it becomes every bird for itself while they stock up on food in preparation for their long southward journey. In addition to bidding their young farewell, the adults will not only be dissolving their pair bond but also be actively moulting with a degree of urgency as the primary feathers must be replaced before migration. Aiming to leave for their African wintering quarters in August, it's thought that the entire wing moult of adult Nightingales may take no longer than 30 to 35 days. This contrasts sharply with resident British Robins, for example, which due to their more sedentary lifestyle can afford the luxury of replacing their flight feathers at double the amount time taken by the Nightingales. So skulking are Nightingales at this time that there are very few records of adult birds observed in the hand after the breeding season, but certainly one individual examined on 9 July had all of its first five primaries in active moult, with none of the replacement feathers being more than a third of their full length. Moulting at such a pace must have meant that this particular bird will have been little more than flightless at this time.

Having begun to moult even before their young fledged, the far more protracted feather replacement of adult Blue Tits will see their plumage not looking fresh again until at least September. The timing of Blue Tit breeding attempts of course varies enormously according to location, so while northern Blue Tits are still likely to be feeding young in the nest in late June, those young which fledged from southern Britain earlier in the month will already be severing the parental ties. Having successfully graduated from their two-week course in finding food and avoiding being eaten, the youngsters will now have to put what they learnt into practice!

Certainly in the southern half of Britain many first broods of Swallows should now be at the point of fledging. As few nests fail completely, Barn Swallows usually have a healthy breeding record, with 90% of eggs laid going on to hatch and a further 80 to 90% of those hatching successfully managing to fledge. This means that a nest of four or five fully-feathered young, busily stretching, flapping and preening, can quickly become a hot and crowded place as departure day approaches. As the young prepare to leave, both the number and size of feeds by their parents are thought to slightly drop in order to encourage the birds out of the nest and onto a nearby roof, tree branch or overhead wire. Exactly when the young fledge depends on both the ability of their parents and the abundance of food, but tends to vary anywhere between 18 and 23 days, with an average time settling at around three weeks. Once one fledgling makes its bid for freedom, it tends to embolden the rest to leave too, with all the young usually departing on the same day. Certainly the first week of liberation will see the young staying close to the immediate vicinity of the nest site, while keeping away from any other Swallow families also busily fledging. Here the youngsters' strategy is to initially remain quietly perching until one of their parents approaches with food, whereupon they will instantly turn into full begging mode by fluttering their wings and calling vociferously to attract attention. Any Swallow youngsters partly obscured at this stage will need to ensure their calls are loud enough for the adults to find them.

After just a few days the youngsters will have gained sufficient confidence in the air to see them flying up to meet the parents, to receive their food quotas. As they learn to catch insects for themselves, many flight trainees will also hone their flying skills by using objects like grass stems or feathers, which are frequently dropped, before then being re-caught on the wing. Even a week after fledging, the young may return to the building where they nested, the vegetation close by or even the nest itself to roost each night, by which time they should also have mastered the ability of finding and catching enough food to keep them sufficiently well fed. Only when reaching this point of independence will the youngsters then begin mixing with any of the other Swallows from neighbouring nests as the family structure slowly begins to break down. Relieved to be in a position to abdicate any responsibility towards their first brood will mean the adults can then turn their attention towards the laying of a second clutch. In a study of Danish Swallows, 70% of pairs that went on to lay a second clutch used exactly the same nest, and new nests were only more likely to be constructed in those cases where an infestation of mites had built up to such a level that it would have heavily impacted on the well-being of the chicks.

As the Swallows are preparing to lay their second brood, the short incubation period and quick fledging times of British Robins should mean that by late June many established pairs will already have fledged their second brood. It's thought that the majority of Robin pairs will only attempt a third clutch in an exceptional year, with most settling instead for just the two broods. Stopping at two will mean the female has no need to sidle away either to make a new nest or spring clean one of her previous efforts, but will enable her instead to be on hand to ensure as many youngsters as possible are carefully steered through to the point of independence in early July.

The end of June should also see any Kingfisher pairs that attempted to raise a second brood midway through their second incubation stint of the year. With the fledged youngsters from the first brood long since having been chased out of their parents' territory, a quiet normality will have temporarily returned to the riverbank as the female once again carries out her sitting vigil. Brief respite from the dark monotony of the nest chamber will only come when she is called out by her mate for a stretch and an opportunity to catch a fish or two, before then disappearing back up the tunnel before her eggs begin to chill.

Unlike Kingfishers, where the adults drive away their newly fledged offspring no more than four or five days after leaving the sanctuary of their bank-side nest, Tawny Owl parents are willing to cut their delinquent offspring considerably more slack when it comes to ushering them towards self-sufficiency. Despite having climbed out of the nest cavity as early as May, the youngsters will in all probability still not yet have caught a single item for themselves, as they relentlessly beseech their parents to keep up the constant supply of food. In all likelihood it will be at least another month before these slow-coaches finally realise the only way to get ahead is by taking matters into their own talons. Despite flecks of down still being present around the facial disc and across the wing coverts, giving them a 'frosty' look, the young will now begin to resemble their parents much more closely with each passing week.

Like the Tawny Owls, Peregrine Falcons are another species willing to give their young plenty of time to learn the complex array of techniques needed for hunting before the apron strings are finally cut later in the year. The Peregrine adolescents also seem much keener students than the lazy juvenile Tawnies and by now should have already developed all-important confidence in the air as they begin chasing each other, harrying passing gulls and practising the stoop that will ultimately reap dividends when hunting for themselves. All these exercises might look like play to the uninitiated, but in fact are a vital step in both improving coordination and strengthening their flight muscles. At this stage some of the aerial flight lessons can be amazing to watch, particularly when the adults fly in with food. Streaking across the sky, the adult can often have all the screaming youngsters on its tail, with the successful juvenile either taking the prize by flipping underneath its parent to grasp the prey straight from the talons or catching the food in mid-air after being dropped by its diligent parent. As the young grow stronger and more able on the wing, the parents may well then decide to move away from the nest site during the day to escape their incessant begging, only returning either with food or to roost.

With begging practised to a fine art, the Cuckoo chick by late June will be driving its foster parents on to keep it fed, close to the nest from which it fledged, as it works steadily towards independence. The Cuckoo's biological father, however, whose sole contribution to parenthood was impregnating the youngster's biological mother, will by now have either left Britain, or at least be on the verge of departure. Obviously the window in which the females are able to target hosts' nests is quite a small one, and so by late June, as the females target the last of any late clutches, the redundant males will have little qualm in leaving the females and chicks to their own devices.

Certainly Chris the Cuckoo, satellite tracked by the BTO from when he was initially tagged as a one-year-old bird in 2011 until his untimely death in the summer of 2015, was recorded as leaving for the continent anywhere between 17 June and 7 July. This suggests that most male Cuckoos rarely spend more than seven weeks – or a mere 14% of their year – in Britain, before then crossing the English Channel for all points south. From the data collected from Chris and the other tracked Cuckoos, it seems that upon leaving Britain they are in no hurry to move south too rapidly and often appear quite happy to feed up in north-west Europe for at least a week before turning towards the Mediterranean. June 2012, for example, was a very eventful time in Chris's life, as he bucked the trend of other years by taking in a European tour which started in Belgium, before then visiting Germany, Luxembourg and the Netherlands in quick succession, only to end the month back in north Belgium, near the docks at Antwerp! Despite no females having yet been tracked, as their marginally lighter weight still prevents the current version of satellite transmitters from being attached, there is currently no reason to suggest that the females will do anything radically different to the males when they too leave a short while later.

Safely ensconced underground, once the Puffin chick reaches around a week old, it should be able to maintain its body temperature well enough so that it can be left unsupervised while both parents are out fishing. The adults never bring food in at night, and there is also a very pronounced rhythm to when fish are delivered, with a high frequency of feeds in the morning, a tailing off during the middle of the day and a resurgence again in the evening. These pronounced peaks and troughs may well be due to the behaviour of the fish that are being exploited by the Puffins, with Sandeels more actively feeding in the water column in the morning and fish like Sprats rising closer to the surface at dusk. Once the morning's fishing gets under way, the traffic of birds flying to and fro can be very clearly seen at large puffinries, with the adults frequently returning like geese in a 'V' formation to deliver sustenance to their waiting chicks. Once the food has been dropped off, the parent will often not fly back out to sea immediately, but use this brief period of respite to glean from other Puffins valuable information, not just about the directions leading to the best fishing areas, but also the type and quantity of food being brought in.

The frequency of feeds during the course of the day will depend on how far the parents are foraging out to sea and the availability of food, but generally most chicks will receive anywhere between three and eleven feeds in a day, with four to six mealtimes being most common. When younger, the Puffling would have been carefully fed beak to beak by its parents, but as it develops, the parents may confine visits to simply dumping the catch on the floor of the burrow and leaving the youngster to waddle down in order to feed itself. The chicks are able to put away a surprisingly large amount of food, and seem to have no problem in polishing off every last morsel brought in by their parents. When not eating or sleeping, the growing Puffling will also become much more mobile underground and spend an increasing amount of time preening, scratching and playing with any nesting material close by. The period when the young hatch also seems to coincide with a small window of time when a few of the two- and three-year-olds from previous breeding seasons pay the colony a visit. Too young to breed and presumably dropping in just to conduct a reconnaissance, these adolescents with their dusky faces and thin bills will often stand out like sore thumbs amongst the throng of adults in all their breeding finery.

As the teenage Puffins familiarise themselves with a thriving puffinry, finally towards the end of the month the female Waxwings breeding up in the taiga forests of northern Scandinavia and Russia should see the first chicks chipping out of their eggs after around a fortnight's incubation. The young, like many perching birds, hatch naked and blind, and so will initially be totally dependent on their parents for their every need. Due to the remote locations where Waxwings breed, relatively few data have been collected on their breeding habits, but a study in southern Norway revealed that the young were still brooded by the female for long periods even at five days old. While the female broods the chicks, the male's responsibility will be to forage for food. Mostly insect food is considered to be brought in during the first few days and the regurgitated bolus of insects will either be passed to his mate to feed the chicks, or fed directly to each expectant red gape in turn.

Even further north on the Russian Arctic tundra, those Bewick's Swans which both laid claim to a territory and built a nest should by now be well into their 30-day incubation period. Despite the vast majority of swans making the journey all the way back to the breeding grounds, for one reason or another a large proportion may have already made the decision not to breed. Without the stresses and strains of incubating and rearing cygnets, these non-breeding birds will have plenty of time to feed in the perpetual daylight experienced during the summer solstice north of the Arctic Circle. When not feeding or sleeping, the young swans will also probably be making use of any spare time to practise the art of courtship in the hope of having more luck the following year.

July

Marginally shading August for the mantle of 'warmest month', the colour purple often holds sway across town and country in July as flowers like Buddleia, Heather and Knapweed all bloom profusely to coincide with the super-abundance of insects emerging in the long, warm days of Britain in mid-summer. While a whole plethora of butterflies and dragonflies take to the wing to make the most of their short lives, many birds, however, will be keeping a far lower profile. With courtship and territorial songs suddenly replaced by subdued contact calls and the sound of incessant begging from young still in the nest, for the avian fraternity July is primarily a time for feeding, fledging and moulting.

Early July

Finally, after just over a month of incessant incubation in the perpetual daylight of the Arctic tundra, those breeding Bewick's Swans that managed to keep a whole host of marauding predators away from their clutch should be finally welcoming their newly hatched chicks into the world. It seems the hatching process can be a protracted one, with the chicks emerging over a period lasting anywhere between one and three days. From research carried out on the breeding grounds it seems that around 90% of the eggs will hatch, while the other 10% were either infertile or the embryo died during development. Throughout this key period both parents will be close at hand to ensure any unwelcome visitors are kept at bay and once dried off, the downy young will be fully sighted and able to feed themselves within hours of emerging into the maritime tundra.

Also able to rear just a single brood during the brief northern Scandinavian summer, most Waxwings should by now be busily feeding their young at the beginning of what is statistically the warmest month of the year in the taiga forests. Needing to brood the grey-brown chicks much less from around day six means the female will suddenly become freed up to help share the chore of finding enough food during what will be the youngsters' phase of quickest growth and most rapid development. After being fed initially on protein-rich insects such as mosquitoes and midges, the parents will then subsequently feed their noisy young on a combination of invertebrates and locally foraged fruit. Certainly in the Swedish taiga forests the young Waxwings will probably be fed a combination of Crowberry, Bilberry,

Cowberry and Bearberry. Having flowered in May and June, most of these plants should be forming their clusters of fresh berries by the time the Waxwing chicks are at their most ravenous. Further east in the Russian Federation, the later arrival of spring might just mean the nestlings have to settle with any berries still remaining from the previous year's stock, in addition to a healthy supply of flies, beetles, flying ants, caddisflies and other insects.

Compared to Waxwings, the Puffin parents will have to make far longer foraging trips during the breeding season if they are to locate sufficient food for their single chick back on dry land. How far the Puffins travel to catch fish for their subterranean Puffling was, until recently, little more than educated guesswork. Research work carried out on the large Skomer colony, in west Wales, found 85% of Puffins recorded within 15km of the island were carrying fish out at sea, meaning these must have been birds returning back with food for their chicks. However, on the remote islands of St Kilda off west Scotland, the researchers had to travel 40km from the islands before the majority of Puffins were seen to be carrying fish, suggesting that the birds may have to forage at greater distance. More recently seabird researchers Mike Harris and Sarah Wanless have successfully managed to attach GPS (Global Positioning System) loggers to a number of Puffins on the Isle of May, situated at the mouth of the Firth of Forth in east Scotland, and found that the average distance the Puffins were travelling to feed was 38km, with a maximum recorded of 64km – a long way indeed for a fish supper!

Fish are of course never evenly distributed across the seas and oceans, but distinctly clumped, and so finding the discrete shoals of fish in the vast open spaces of the North Sea, Irish Sea and north-east Atlantic Ocean could potentially be very difficult. It seems, however, that Puffins, through trial and error, must be able to quickly familiarise themselves with the best feeding areas. Despite being intensely sociable birds on dry land, Puffins seem to shun company at sea by keeping spaced out across any possible fishing areas. It's believed that each Puffin dive will rarely last longer than a minute, with the bird then taking a brief rest on the surface before slipping below again. Using time-depth recorders attached to Puffins, Sarah Wanless and her colleagues recorded one Puffin making an astonishing 194 dives in 84 minutes, which suggests that the Puffins don't dive anywhere near as deep as Guillemots and Razorbills, for example. Below the water Puffins suddenly become transformed into lean, mean swimming machines as the pressure compresses their feathers flat to the body and the wings are suddenly transformed into powerful flippers able to propel the bird through the water at a surprising speed. With the feet acting as both rudder and stabilisers, their manoeuvrability is another tool in the armoury helping them to snatch any unwary prey.

Further results from four time-depth recorders attached to Puffins feeding chicks revealed that, on average, each bird made 1,148 dives per day and spent 7.8 hours under water. From this revelatory data it has been extrapolated that Puffins may catch a fish on only two out of every five dives, and so energy-demanding is this occupation that 90% of all the fish they catch will be purely for their own consumption, with no more than 10% destined for the chick waiting back in its burrow! Of course, unlike the Blue Tits feeding their chicks, which will return to feed their young with a single caterpillar each time, Puffins are famously able to carry many fish at a time. This ability to catch more fish while still holding on to the ones already caught is achieved with backward-pointing spines, called denticles, on the roof of the mouth. Adopting the 'belt and braces technique', any food caught is prevented from being dropped by a rough covering at the back of the tongue, and finally the load they're able to carry is also maximised with the help of a flexible hinge between the lower and upper mandibles.

With Sandeels being the main prey for British Puffins, the number they're able to carry will vary between colonies, different pairs and from year to year. Billfuls of between four and twelve seem the norm at many colonies, but an unbelievable 61 Sandeels and a Rockling observed being carried by one bird returning to feed its chick on St Kilda is frankly astonishing. Historically, there was a suggestion that the Puffins arranged the fish to be positioned alternately in the bill, so that the head of one fish would be placed adjacent to the tail of the next and so on, but in reality the fish are carried in a haphazard fashion and presumably held in the same orientation as when the fish was originally caught. However they're arranged, once the adult is content it has caught a sufficient number, it will then head back to the puffinry, where it must face one last challenge before the chick receives the reward its patience deserves.

Having now hatched around twelve weeks ago, the juvenile Tawny Owls, like the Pufflings, will still be reliant on their parents for providing 100% of their dietary needs. Despite being fully mobile, fledged young from the same nest will still frequently choose to roost close together at a few favoured spots as they wait for their parents to bring in food. It seems the parental care that the adult Tawnies provide is confined to the provision of food and defence against predators, with little evidence that they actually train their young how to hunt or catch prey. This is a different ethos to Peregrine Falcon parents, which take the tutoring of their offspring very seriously when it comes to passing on best hunting practices. Once their young have begun to master the art of aerial food-passes, the parents will then move on to delivering live birds. These usually tend to be feral pigeons that the parents have already caught alive, only to be released, possibly in a dazed and confused state, as catching practice for the young. Initially there are plenty of near-misses as the quarry either escapes or has to be subsequently recaptured and killed by the parents, but the youngsters will soon become proficient at catching and dispatching the prey item with the minimum of fuss.

As the young Blue Tits begin to disperse further afield from where they were reared, they will not be figuring prominently on any Peregrine Falcon's radar, but much higher on the list of any local Sparrowhawk, which would jump at the chance of an easy meal of a 'naive fledgling' or two. In addition to running the very real risks of predation, the number of caterpillars in the trees may have already passed its peak too, leading to starvation and any diseases brought on by being in poor condition, which can be major stumbling blocks during this seeming 'time of plenty'. However, there is substantial help for the many Blue Tits which fledge from suburban locations, thanks to easy feeding opportunities offered up in many gardens. Certainly in the south of Britain, most gardeners who regularly leave out peanuts and sunflower hearts can be sure of seeing a steady stream of juvenile Blue Tits in the early part of the month. Distinctive with their greenish caps and yellowish cheeks, many juveniles can now be seen dashing from the cover of the surrounding bushes to the feeders as they attempt to grab an easy beakful of food.

Any adult Blue Tits freed from the constraints of looking after their young should by now be able to concentrate on finding sufficient food to fuel their annual moult, which could carry on at least up to early September. Of course, adult Nightingales will not have the luxury of being able to carry out their wing moult in such a gradual fashion as their flight feathers will be needed to power them to Africa, and so a rapid moult will make them far less mobile at this time. Needing to keep flight to a minimum, it's thought that the adults will choose sites away from the breeding grounds, which offer both rich feeding opportunities and plenty of places to hide away from any predators during this perilous period.

Having also deserted their breeding grounds, the adult Lapwings and any fledged juveniles will by now have been absorbed into large, mobile moulting flocks. As good feeding opportunities for these flocks in summer may be both widely spaced apart in the countryside and ephemeral by nature, the Lapwings will need to ensure they're able to stay aerial. This of course means – like with the Blue Tits – that the wing moult is a long, drawn-out process, not finishing until early September. Also needing to be prepared for the rigours of winter ahead, the juveniles will begin a partial moult by shedding some of their fledging feathers. Clearly distinct with their short crest, buff face, scalloped back and wing feathers and incomplete breast-band, the juveniles will also show a shorter wing than fully adult birds with a clearly narrow primary area in flight. It will not be until December, and some seven months after hatching, that they will finally be looking pretty similar to the parents that successfully raised them earlier in the year.

Early July should also see the second brood of fledged Robins receiving their final life lessons before branching out on their own. Providing their parents haven't opted for a third brood, the chicks from this second clutch will have had the benefit of both parents' attention in helping briefly to cushion them from the harsh realities of life while they learn the skills necessary for surviving on their own. After just three weeks following their parents' lead, the young are considered as ready as ever, and with little or no ceremony simply disperse into the countryside to prepare for the difficult times ahead. The departure of the young will trigger a number of changes in the adults. Firstly the pair will kick-start their annual moult, a slow process which may well take up to eight weeks as feathers worn down by the rigours of the breeding season, exposure to ultraviolet light and the attentions of parasites are replaced. During the moult, the strict territoriality which had been in place since the previous autumn will also break down. Not needing to hold a territory will render the Robin's song redundant with the knock-on effect that July and August are the only months of the year when town and country are devoid of their beautiful songs. In contrast to the rest of the year, when Robins will be both loud and visible, this vulnerable period will find them keeping mostly hidden away in the undergrowth as they concentrate on finding sufficient food to power the growth of their replacement feathers. Becoming antisocial once again, any remaining pair bonds quickly dissolve as each bird returns to splendid isolation.

While the adult Robins retire from public view it's quite a different scenario for any Swallows intent on laying their second clutch. For those pairs successfully raising a first brood, it's highly likely that they'll have an 'if it ain't broke, don't fix it' attitude towards their relationship and will stay together for the rest of the breeding season. But matrimonial strife, brought on by the first clutch failing, can often result in the pair opting instead for a quick divorce to try their luck elsewhere. Clutch size is thought to be lower for second broods as the amount of daylight available for feeding slowly declines, but the number of eggs can also be influenced by the condition and experience of the female. When the clutch is completed, it will once again be the female's responsibility to incubate the eggs through to hatching while the male bides his time in preparation for the new arrivals.

Any fish observed being carried back to the nest-bank and down the tunnel will be a sure-fire sign that the second clutch of Kingfisher chicks will have finally hatched out in their dark and gloomy, but well protected, chamber. The first small fish initially brought in will then be carefully placed head first down the gape of the naked and blind chicks as they digest the first of many similar meals over the course of the following 24 or 25 days.

Having already departed our shores on their long migration back to the Congolian rainforests, by early July the male British-breeding Cuckoos will have dispersed in a wide arc across Europe. Despite the BTO's satellite-tracked Cuckoos being reported at this stage from a whole host of countries right across Europe, there are already early signs indicating a split between those Cuckoos that are opting for the more westerly flyway through Spain, and other individuals that are planning to take the far more central European migratory route through Germany and Italy. Irrespective of the direction chosen, presumably experience will also play a key part in the birds' survival as any seasoned campaigners link up with those rich feeding sites familiar from previous migrations. The Cuckoos' mission at this time of year is quite simple – to stock up as well and as quickly as possible before heading south across the Mediterranean Sea, crossing most, if not all, of the Sahara Desert in one fell swoop.

Thousands of kilometres away in Britain, and after around 17 days spent dining in its host's nest, and a similar period begging from its hoodwinked foster parents, the Cuckoo youngster should by now be on the threshold of independence. Leaving its hosts to rescue what remains of their breeding season, the young Cuckoo will then begin seeking out hairy caterpillars, which also happened to be the food of choice of its parents earlier in the summer. With the young Cuckoos only just learning to feed themselves, it's remarkable to think that their errant parents will have already departed for the continent, leaving the chicks they never met to begin their long migration alone down to their wintering grounds in the heart of Africa.

Mid-July

After little more than two weeks on a diet of berries and mashed up insects, the young Waxwings will be already be raring to sample the delights of the taiga in summertime. They will fledge looking remarkably inconspicuous and with only a hint of the crest, black bandit's mask and pinkish coloration so distinctive of their parents. Fluttering out one by one, the young are initially barely able to fly, and so will need to stay close to their parents for a further two weeks until the power of flight has been mastered, they have learnt the best locations for feeding and become proficient in the most effective techniques for catching insects. Sitting almost astride the Arctic Circle, the city of Rovaniemi is the commercial capital of Finland's northernmost province, Lapland, and will certainly have a healthy population of Waxwings breeding in its surrounding forests. At this time of year, the average daily temperature in Rovaniemi will reach an annual high of around 15°C with daylight virtually around the clock. These conditions should enable Waxwing families to have plenty of time and opportunity to feed well before the slow and inexorable descent towards winter eventually forces them off their breeding grounds and to a more benign climate further south.

Even further north than Rovaniemi, on the Russian tundra those Bewick's Swans with recently hatched chicks will be keen for their brood to leave the nest for the relative sanctuary of nearby water as soon as possible. The instant the family unit can become mobile the cygnets' survival prospects will be hugely enhanced, as being on water will instantly offer more protection from most land-based predators, while the broken terrain away from the nest should also offer more opportunities to remain safely hidden were danger to beckon. Unlike Mute Swans, the recently hatched cygnets will not ride on their parents' backs and are still brooded regularly at night and during inclement weather. The parental duties will also be split at this stage, with the job of brooding the young largely falling to the female, while her mate's responsibility is to keep a constant lookout for predators such as Arctic Foxes. If a fox does approach too closely, the male should be able to use his size, bulk and aggression to drive away most unwanted attention, but irrespective of the male's attentiveness, a number of young will inevitably be predated early on. Upon leaving the nest the cygnets are totally capable of feeding themselves; however, when very young they will still have to rely on at least some help from their parents with the collection of any food simply out of reach. Many bird species will dissolve the family ties not long after their young fledge, but this isn't the case with Bewick's Swans, as the bond will only continue to grow stronger between the young and their parents as the season progresses. In fact, so pronounced is this familial affiliation that those youngsters successfully managing to fledge from the tundra will, in all likelihood, stay together with their parents through the subsequent winter right up to spring the following year.

Fed on a constant stream of fish, the Puffling will have grown considerably since it hatched around a month ago and will also have managed to shed much of its down to reveal a sleek plumage not dissimilar to its parents. Its upper surface should by now be a matt black, with a clean white underside, and charcoal-coloured cheeks topped off by a thin, dark bill. It will also have upped its exercise regime in preparation for its life at sea, as countless hours will now be devoted to preening, scratching and flapping. When close to fledging the Pufflings will suddenly begin venturing to the burrow's entrance in the evening to exercise their wings and learn the visual cues they will need if and when they return as adults. On these brief forays which will take them to the very threshold of the big, wide world they need to be incredibly wary of any lurking predators keen for an easy, defenceless meal, and so look almost nervous before disappearing back down below to the sanctuary of the burrow.

It's not just the Pufflings that need to be careful, as the adults returning with food also risk being attacked by other seabirds, which either specialise in stealing food, or killing and eating the Puffins themselves. Great Black-backed Gulls and Great Skuas (or Bonxies), for example, are easily capable of overpowering and killing any Puffin they can catch. When out at sea the Puffins are relatively safe as they're able to escape most dangers by diving below the surface, but returning to the colony presents a much higher risk. While Bonxies use their agile nature to catch Puffins in mid-air, most gulls' techniques seem to consist of little more than catching them unaware as they enter or leave their burrows. At some seabird colonies, the local breeding pair of Peregrine Falcons may also be partial to a Puffin or two when the opportunity arises. The commonest thieves of the food being carried back by the Puffins to feed their chicks tend to be Arctic Skuas, Herring Gulls and Lesser Black-backed Gulls, but at many of the northern colonies it is the Arctic Skua that has kleptoparasitism down to a fine art. Being sleek and streamlined, with long pointed wings and a lightweight body, the Arctic Skua has become perfectly evolved for chasing fish-laden Puffins. Able to roll, turn and dive at great speed, the skuas can change both their height in a split second, and direction on a sixpence, as they harass any returning Puffins into dropping their load. Once spilt, the ill-gotten gains are then acrobatically caught before either being lost in the water below or intercepted by less manoeuvrable predators keen for an easy meal.

The best way for the Puffins to reduce this risk of attack is to join forces with other fish-carrying birds. Linking up to form aerial flocks out at sea, the final assault on to land can then be made in the safety of numbers. Puffin wheels can also be deployed in an attempt to confuse any predators from fastening on to any one target from amongst the swirling flock, thus helping the Puffins to get to and from the burrows in a safe and highly synchronised manner. From studies carried out at different puffinries the amount of food lost to skuas and gulls is rarely disastrous. For example, on the island of Foula, off Shetland, fewer than 1% of the Puffins were chased, with only one in five forced to drop their fish, while on the Isle of May, off the east coast of Scotland, 5% of adults lost their food to gulls. Irrespective of the levels of theft, providing there are healthy stocks of Sandeels, the adult Puffins should be able to ride the occasional loss without their chick going unduly hungry.

Back at the Peregrine's nest site, the parents will be doing their level best to ensure their adolescent young are kept well fed while learning the necessary skills to fend for themselves later in the year. The young will have been conducting mock attacks on other bird species flying past the nest site for at least a few weeks, and at some point these start to get serious as the juveniles attempt to make a kill. If their targets are skilful flyers, such as pigeons, the young Peregrines may well be repeatedly unsuccessful, but at some point they will strike it lucky by both catching and dispatching their very first meal.

Having done very little apart from avoiding being predated and constantly harassing their parents for food, the Tawny Owl chicks will soon be staring into the abyss of venturing out on their own without the life support system of their parents to keep them fed and safe. With just a few flecks of down still around their faces, the youngsters should be looking pretty similar to their parents by now, and Tawny Owl expert Dave Culley believes any siblings that branched together will at this stage still be spending most of their time in each other's company. Dave believes, certainly for his suburban Tawny Owls, that slugs form a substantial part of the diet at this time of year and has observed youngsters watching their parents catch them on the ground. Technique duly noted, the juveniles may then proceed to copy the same foraging behaviour as they take their first tentative steps towards self-sufficiency.

Reaching the point of independence somewhat earlier than both Peregrine and Tawny Owl young, as the Swallow juveniles from first broods are now able to feed themselves completely unaided, this seems to precipitate a movement away from the site where they were reared. Still vulnerable to predation from birds such as Hobbies, the young often tend to scatter into the surrounding countryside, doubtless to find good feeding areas and familiarise themselves with any local landmarks, in preparation for a return the following spring. With the young just as likely to go north as south at this stage, this dispersal does seem to be more a case of random wandering than the beginning of a defined migration to South Africa, which isn't likely to begin in earnest for at least another couple of months.

Back feeding young again, the Kingfisher parents will see their second brood still some way off fledging as fish deliveries begin to rise dramatically to cope with the incessant demands of their young chicks. With the adults alternately brooding the young for at least the first week, the non-sitting bird will be delegated fishing duties, which should be none too onerous given the sheer amount of food on offer at this time of year. Irrespective of the quantity of fish available, they do still need to be caught, however, and relieving a stream, river or pond of its fish is an art that the Kingfisher has certainly mastered like no other in Britain. Their preferred technique usually begins with a period spent quietly perching on a favoured spot anywhere from a metre to three metres above the water as the bird weighs up its options. The Kingfisher is believed to have excellent eyesight and is aided by two fovea, which are highly sensitive areas on the retina where the image appears particularly sharp. Also utilising special elliptical lenses, the bird is not just able to cope with the reflection from the water's surface, but also refraction – the phenomenon where light is deflected, or bent, when passing between air and water. In the instant before its dive, the bird is thought to instinctively make a three-dimensional calculation of its flight path based on the fish's position.

The dive itself is remarkably quick, with the Kingfisher able to leave its perch, execute the dive and return all within little more than a second. Using its wings to increase the acceleration of the dive, the tail can provide any micro-adjustments before the bird hits the water, by which time it will have covered its eyes with a special nictitating membrane and have its bill already agape. The Kingfisher is not a deep diver, with most fish caught just below the surface, and if it does manage to grab a fish in its bill it will then quickly power out of the water on beating wings, helped by a positive buoyancy due to air trapped in its feathers. The Kingfisher is certainly not a spear fisher, and so any tiny miscalculation that results in the fish being accidentally impaled on either mandible, will usually end with the prey being dropped into the water and lost. However, any fish snaffled in the textbook manner will be quickly dispatched, as the now perching bird whacks the fish's head on a branch, or something equally unforgiving close at hand, to immobilise it. In the case of Sticklebacks the Kingfisher must be careful of the spines, and particularly robust species, like Bullheads, may need some serious blunt force trauma to either kill the fish outright or at least render it sufficiently dazed to be swallowed without causing the birds any harm. If the fish is considered sustenance for the parent itself, it will be quickly swivelled around in the bill before disappearing down the throat head first, but if destined for the brood, the adult will then return straight back to the nest before offering the meal head first to its chick of choice. Repeated and constant immersion in water throughout the day will mean that in addition to the pressure of finding enough food, the adults will also need to spend a large amount of time preening and oiling their feathers from an oil gland on their rump to ensure they don't become waterlogged. Even for those pairs still rearing young, moulting may already have begun as their flight feathers are replaced in a slow, methodical fashion that will last until at least November, or even later in those cases where the breeding season may have been delayed.

With their young dispersed, those adult Robins that limited their breeding season to just two attempts will be busily keeping out of the public eye as they replace their entire plumage. It's vital that they're able to retain the power of flight during this period, and so the feathers will be shed and regrown in a slow, precise and controlled manner. The moult starts in the middle of each wing, with the primary feathers moulting outwards and the secondaries moulting inwards. Usually the first feather to be shed is the innermost primary, with each successive feather then being shed every four or five days, so by the time the fifth primary has been dropped the first two primaries should have completely regrown. Vital for manoeuvrability, the tail feathers are shed from the centre pair outwards and their slow replacement will also span the entire period of the wing moult. Taking around two months, by the time the flight and tail feathers will have been exchanged, the Robin will also have replaced all of its body feathers. Only then, looking immaculate in its fresh plumage, will the Robin be ready to finally show its face again as it prepares to commence battle for the all-important winter territory.

The mass feather replacement that is the annual moult won't just be restricted to Robins at this time, as Blue Tits, Lapwings and Nightingales will also be busily shedding and regrowing feathers in preparation for either a winter spent in Britain or a more benign climate elsewhere. As the adult Blue Tits slowly continue their moult, their young will also begin to replace the hastily grown juvenile feathers for a set that will allow them to survive their first winter and last until their first breeding season. For these immature birds the most important feathers to replace will be those around the head and body, which will not just ensure better insulation but enable them to play a full and vigorous part in the mating game the following spring. This post-juvenile moult means that young Blue Tits will not actually replace their flight and tail feathers until completing a full annual moult following their first breeding season, resulting in them only becoming indistinguishable from fully adult Blue Tits at the grand old age of 17 months.

Also keen to retain the power of flight during their moult, the annual replacement of feathers will not prevent a large number of continental Lapwings from reaching Britain throughout July and August. Arriving first on the east coast, these Lapwings which will have summered on the continent will probably just slowly melt into the flocks of British-breeding Lapwings already roving the countryside. The origin of these Lapwings may vary from eastern European birds, which commonly undertake a post-breeding westerly or north-westerly movement to reach Britain, to those dispersing south-west from breeding grounds in Scandinavia.

By contrast, no such advance movement will be seen from those Nightingales that successfully bred in Britain earlier in the summer as their rapid moult will see them quietly feeding away as they attempt to take on sufficient fuel to power the quick growth of their new feathers before their flight to west Africa. Like the Blue Tits, the juvenile Nightingales will also at this stage be undergoing a partial moult of primarily their body feathers and some of the coverts (the feathers covering the main flight feathers) but their primaries will not be replaced until this time next year.

By the middle of July, data collected by the BTO suggests that the male Cuckoos will have already made their choice as to the migratory route they will take into Africa, with the birds seemingly split between those Cuckoos preferring the Iberian route and those deciding to travel via Italy. Chris the Cuckoo is currently the only bird to have successfully carried a fully functioning transmitter for four years until his demise in the summer of 2015 and, interestingly, was a bird that always chose the Italian route down to Africa. It seems certain that Chris, and possibly many of the other Cuckoos too, tend to be site-faithful to a few tried and tested locations during their migratory route and undoubtedly for Chris the most important stopover in southern Europe was the huge Po Valley in northern Italy. Travelling over 600km from its source in the Cottian Alps to its mouth in the Adriatic Sea, the River Po is the longest in Italy and surrounded by an expansive, fertile flood plain, which has now largely been converted to agriculture. The delta on the Adriatic coast is considered particularly important for migratory birds with the Po Delta Regional Park designated as a World Heritage Site in 1999. Consisting of wetlands, forest, dunes and salt pans, this site must obviously be the perfect place for a British-breeding Cuckoo to spend time fattening up before traversing both the Mediterranean and the Sahara Desert in one enormous leap.

With their biological parents ready to exchange Europe for Africa, the young Cuckoos should still be feeding up in Britain prior to their own departure. After fledging there appears to be a period where the young Cuckoos will leave the location where they were reared to spread out in any direction. This seemingly random movement mostly tends to occur over a short distance, and despite British juvenile Cuckoos having been recorded as far away as Denmark and Germany, this 'wanderlust' shouldn't be confused with their proper migration to Africa, which will still not be conducted for a couple of months. In many ways the young Cuckoos' journey is even more remarkable than that of their errant parents, as they will not have been able to either learn the route from older, more experienced birds or orientate themselves with the help of familiar features in the countryside. The only conclusion to draw is that the young must follow some kind of genetically hard-wired satellite navigation system, which not only specifies both direction and distance but can also be re-calibrated for variable weather conditions or any obstacles encountered along the way. Irrespective of how this feat is achieved, their piloting skills across thousands of kilometres of utterly alien territory remains one of the great feats of the natural world!

Late July

After six weeks spent underground being waited on 'wing and foot' by its parents, the young Puffling should by now be finally ready to leave the confines of the burrow to make its bid for freedom. Weighing in at around 35 to 40g when hatching, the Puffling will have reached between 250 and 350g by week five, before this final week sees the bird frequently losing a little weight. This 'slimming' programme comes about as a result of a rapid increase in exercise from the constant flapping of its wings and also a decrease in the amount of food brought in by the parents. This crucial weight reduction will primarily ensure the young Puffin is not too obese to 'lift off' at the defining moment it decides to exchange the land for the sea. Despite losing around 10% of its body weight during this time, the Puffling will still have retained more than enough reserves to see it through those first few critical days out at sea while honing its foraging skills.

With so many avian predators keen to make a meal out of a naive young Puffling, the youngsters will only ever leave for the surrounding sea under the cover of darkness. Although the fledging of the youngsters across a large colony can often be spread over a few weeks, on a good night hundreds will often leave the colony within a few hours. Moving to the burrow entrance, the Puffling will often linger at the entrance to get its bearings before a last minute preen and a fluffing-up of its feathers then sees it take to the air for the first time. A good sense of direction is vital as the Puffling attempts the most important short flight it will ever make, often only staying airborne just long enough to splash down in the comparative safety of the water, no more than a few hundred metres from where it was reared. Having successfully navigated away from the puffinry and any chick-hunting gulls still patrolling the burrows on a late shift, the young Puffin will take to the sea just like a duck to water. Able to swim and dive instantly and keen to capitalise on such a confident start, it will then set about distancing itself from all the predators circling the colony to reach the safer waters further offshore. Such is the youngster's eagerness to leave its birthplace behind, by daybreak it could have already paddled several miles out to sea, perhaps even linking up with other recently fledged Pufflings or any adult birds loafing around at the surface in preparation for their post-breeding moult. The parents are thought to play no active part in encouraging the young from the burrows, to the extent that some adults may even briefly continue to bring back food to an empty burrow, seemingly unaware that their chick has already flown the coop. With their chicks having fledged there will then be little incentive for the adults to linger any longer than necessary, and a rapid desertion of the colony will quickly ensue, leaving the puffinry pretty much devoid of life by the end of the month.

Able to feed themselves almost immediately upon leaving their burrows, the juvenile Puffins are certainly quicker learners than the Tawny Owls which, despite having clambered out of their nest as long ago as early May, will only now begin hunting their own food in earnest. Whether the young birds are pushed or jump towards self-sufficiency is unclear, but the motivation may well be a combination of the parents slowly withdrawing their 'meals on wings' service as the youngsters themselves steadily take matters into their own hands. Either way, as the incessant begging gradually declines the young will probably only be tolerated for a couple more weeks in their parents' territory, before it's made perfectly clear that they'll need to take their chances elsewhere.

Still totally reliant on their parents, the Bewick's Swan cygnets will require a further six weeks before the ability to take to the air should see their chances of survival boosted substantially. Until then, the perennial twin threats of predation and poor weather will continue to exact a heavy toll, and the number of chicks in most broods will continue to decline throughout the month. Even for the most experienced parents, to have any more than two cygnets still alive at this stage of the breeding season must be considered little short of exceptional. Like many birds, adult Bewick's carry out a complete moult every year, but as the swans will need to replace the key flight feathers as quickly as possible before their autumn migration, this will necessitate a period of flightlessness often lasting around three weeks. Obviously during this period of intense moulting, not only are the adults themselves more vulnerable to predation, but they will also be far less able to protect their young. Incredibly, to counter this disadvantage, any pairs still with cygnets are able to stagger their moults. With the female thought to moult first, this ingenious mechanism will mean at least one parent is fully winged for longer, which will in turn give the pair a higher probability of being able to successfully defend their cygnets. On the breeding grounds, the last feathers to be replaced before migration will be those belonging to the tail. The body moult, by contrast, will proceed at a much more leisurely rate, meaning that many body feathers are still actively being replaced well after the swans have arrived in their winter quarters.

For those Bewick's Swan pairs either unsuccessful or simply unwilling to breed, a moult will also have to be completed, but being unencumbered by young means there will be no need to stagger their flightlessness. As holding a territory will also have become an unnecessary luxury for these birds, many will tend to form large moulting flocks away from the tundra and in the comparative safety of shallow sea bays and coastal regions. Carefully chosen to limit the access of a variety of terrestrial predators, these locations will still need to provide plenty of aquatic vegetation for the swans to feed on. Moulting flocks might not be the choice of all the swans without young, however, as some will prefer to remain on the tundra during this vulnerable period. Without the water as a protective barrier, these swans will instead choose to either hide or break into a surprisingly fast run should they spot any predators on the horizon!

In the taiga forests to the south of the Arctic tundra those Waxwings that successfully bred should by now have formed into family parties as the newly fledged youngsters join their parents on foraging missions. Any element of territoriality which existed around the nest will have disappeared the instant the youngsters fledge, as the families roam the forests for food, often over a far wider area than on their wintering grounds. As the short breeding season draws to a close and darkness at night begins to reappear, the dropping temperatures will also see the individual family members begin roosting together to conserve heat. Staying as part of a family unit during this period will also provide the birds with more pairs of eyes to keep a lookout for Sparrowhawks and Merlins, two predators specialising in hunting small, unwary and inexperienced birds. Any excited or alarmed Waxwings that may have spotted potential trouble will quickly adopt an erect posture, with crest raised, to ensure that the other birds in the party are also aware of any imminent danger.

Due to the longer summer encountered back in temperate Britain, those pairs of Kingfishers opting for another brood will see their second clutch of chicks growing quickly as the month draws to a close. Due to the chicks' size there will be little opportunity for the parents to turn around in the crowded nesting chamber, resulting in them having to reverse out of the tunnel after the delivery of yet another meal. When the chicks were young the adults would take great care in gently placing the fish down their throats to ensure they didn't choke. However with their offspring approaching fledging, feeding visits for the parents will consist of little more than a trudge down a slimy tunnel, followed by a slam-dunk down one of the youngster's throats and a reversal back out into the fresh air. At two to three weeks old, the chicks often line up side by side, with their bills pointing towards the tunnel as they await another delivery from their hardworking parents.

By the time that the second brood approaches fledging, those juveniles still going strong from the earlier brood should already have begun to undergo their first moult.

Unlike their parents, currently undergoing a total refit, the juveniles will only replace the head and body plumage during their first year, with most of the flight and tail feathers usually being retained until the following spring. This partial moult will see the youngsters ditching their initially dull plumage as they begin to take on the hues and tones of their parents. However, they will still not be able to pass off as fully mature just yet, as their legs won't develop the characteristic orange-red adult coloration until the following spring.

Bringing up the rear, many Swallows' second broods should finally be hatching as July draws to a close. Following the pattern of the first brood, the removal of the eggshells will often be one of the first indications that the young have begun to hatch, and as the female broods the chicks for the first few days the male will once more have to step up to the plate. Upon hatching, the parents will suddenly take the protection of the nest to a whole new level. Any other Swallows caught close to the nest during this time will be subjected to a barrage of alarm calls and predators like cats or rats will be actively mobbed and even struck if they attempt to interfere with the newly-hatched brood.

For both the young Blue Tits and first brood of Robins, late July should see the continuation of their partial moult, as they slowly embrace the plumage and badges of maturity. Any surviving Blue Tit should by now be in the process of gaining the distinctive white face and bright blue cap, while the spotty brown breast of the fledgling Robin will be making way for the famous orange-red breast which it will use for acquiring and holding a territory in just a few weeks. The adults of both species will also be actively moulting, and much of their time will be taken up by finding the extra food to fuel the 'gas-guzzling' replacement of feathers, while still keen to remain largely out of sight. The data analysed from thousands of ringing reports reveals that adult Blue Tits are much heavier at this time of year than during the breeding season, doubtless due to the combination of a far lighter workload without chicks to feed and the relative abundance of food. Certainly for adult Robins, the need to keep a low profile, and the resulting lack of territorial behaviour, will mean that July and possibly early August may well be the only time during the entire year that the British countryside will be devoid of the Robin's song.

Heard here for close to 11 months of the year, the prolific nature of the Robin's song could not be contrasted more sharply with that of the Nightingale, which will have scarcely uttered a syllable for well over a month now. Needing also to complete a much faster moult than our resident birds will mean that by late July many adult Nightingales will look as though they've just been freshly laundered, as they prepare to leave England for Africa. This period before departure is also believed to coincide with a change in the Nightingales' eating habits, as a largely protein-based diet is replaced by one favouring a higher proportion of carbohydrate. As invertebrate prey becomes steadily more difficult to find, the Nightingales will then switch to the abundant amount of fruit suddenly available, blackberries and elderberries being the perfect energy-rich foods with a long, tiring flight just around the corner.

Having already departed our shores well in advance of the Nightingales, the BTO has discovered that late July or early

August seems to be the prime time when many British Cuckoos will leave any established stopover locations in Spain or Italy for their journey to the heart of Africa. Followed with the help of satellite technology between spring 2011 and the summer of 2015, three out of four of Chris the Cuckoo's southerly migrations have him recorded leaving Italy in the second half of July, only to arrive in Africa a few days later. In 2011, for example, a signal from Chris's transmitter reported him to be close to the River Po in northern Italy on 22 July, only then to be picked up in northern Chad, a distance of over 2,600km away, just 55 hours later! As they power over both the Mediterranean Sea and most of the Sahara Desert in one gigantic hop, it now seems likely that Chad may well be a favoured stopover for many of the British-breeding Cuckoos in July and August. As the Cuckoos generally seem to take a more easterly route on their southward migration compared to their journeys north in spring, this would seem to indicate that many of the birds must pass directly over Libya. With a good number touching down in desertified northern Chad, it seems this arid part of the country may in fact be little more than an opportunity to draw breath before the Cuckoos then quickly move to more fertile regions further south. The richer feeding grounds surrounding Lake Chad in the south of the country represent an excellent opportunity to recover any condition lost while crossing North Africa, and mean the Cuckoos are in no hurry to leave this region until they ultimately head for the tropical forests later in the autumn.

As the summer rolls on, any young Peregrines which fledged in the spring should by now have become sleeker, faster and far more adroit at catching their own food. Having served their apprenticeship well, the loosening of the ties with their parents will probably have been initiated by the young who now be regularly feeding well away from the nest site during the day. Still returning to the comfort blanket of their parents' territory at night, these day trips will not just be helping the young birds to find other good feeding areas, but also locate possible overwintering locations and even potential breeding territories. At those sites with an absence of breeding Peregrines, birds may well now begin turning up again at a whole host of sites that have been abandoned since late winter as the whole population becomes far less territorial and much more mobile.

Mobility will also have become a distinguishing feature in the behaviour of both British-breeding Lapwings and any immigrant birds still pouring over from the continent. Ringing records of Lapwings which were known to have bred elsewhere, before then being noted in Britain, seem to suggest that the largest influxes originate from Belgium, the Netherlands, Denmark, Germany, Sweden and Norway, but there have also been some interesting recoveries from elsewhere. In 2003, for example, a Lapwing found dead on Romney Marsh in Kent was revealed to have been ringed as a nestling in northern Lithuania in 1994, a distance of 1,724km from where it was recorded. This record is one of a handful of Lapwings known to have originated from a broad arc encompassing Estonia to the north and Poland further south, and due to the very limited number of birds ringed in these countries, these records may merely represent the tip of an iceberg. All these countries surrounding the Baltic sea have much colder continental winters than Britain, where freezing temperatures are the norm, and due to the Lapwing's inability to extricate food from frozen ground it's perhaps no surprise that a healthy number will move west to the Gulf Stream-warmed climate found in and around the UK.

August

Synonymous with school holidays and filled with the frenzied activity of families trying to enjoy what remains of the British summer, August is a time when the wildlife will be keen to complete breeding cycles before the nights start drawing in. With the bird breeding season largely over for another year, those visiting just for the summer will either have already departed or be busily stocking up on last minute provisions for the flight. Amongst those birds residing here all year, a continued moult and dispersal into the countryside will be the order of the day, giving the garden that empty feeling of a stadium just after a match has been played. Around our coasts the picture could not be more different, however, as waders like Knot, Turnstone and Sanderling exchange their Arctic breeding grounds in their droves for our mudflats - representing just the vanguard of a whole variety of bird species keen to spend the winter in Britain.

Early August

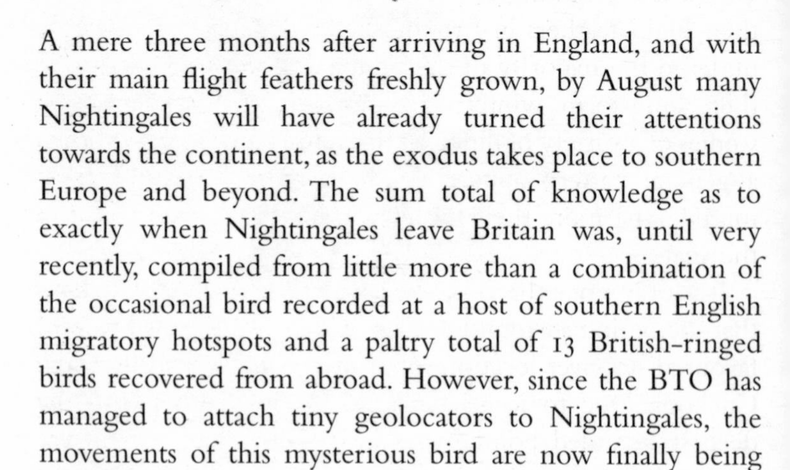

A mere three months after arriving in England, and with their main flight feathers freshly grown, by August many Nightingales will have already turned their attentions towards the continent, as the exodus takes place to southern Europe and beyond. The sum total of knowledge as to exactly when Nightingales leave Britain was, until very recently, compiled from little more than a combination of the occasional bird recorded at a host of southern English migratory hotspots and a paltry total of 13 British-ringed birds recovered from abroad. However, since the BTO has managed to attach tiny geolocators to Nightingales, the movements of this mysterious bird are now finally being revealed.

Of all the Nightingales with geolocators attached in the spring of 2009 and then crucially re-trapped to recover the equipment the following spring, only one individual was able to deliver any meaningful data. The bird, named OAD because of the letters on its device, is believed to have divulged more information in a single stroke about the movements of Nightingales than in a hundred years of ringing. The recovered data revealed that OAD was thought to have left England in early August, only then to be recorded passing west of Paris a week later. Despite a substantial layover in Spain and Portugal, the spring migration north is now believed to be conducted at a much more rapid rate than the southward journey, which appears decidedly far more leisurely by comparison. Needing neither to proclaim a breeding territory nor a mate in Africa, it seems there may not be quite the imperative for the Nightingales to migrate so quickly, thereby leaving them free to feed along the way.

Unlike the Nightingales which will have certainly moulted their flight feathers in Britain, the adult Cuckoos will only undergo the majority of their annual moult upon arrival at their long-term wintering destination. In fact moulting studies of a variety of different migratory species have shown a general pattern appearing; the further south the birds migrate, the more their moult will be delayed until later in the year.

It is only since the BTO began satellite tracking Cuckoos that the region surrounding Lake Chad was revealed to be a favoured stopover location for many of the British-breeding Cuckoos. Due to its particularly shallow nature, with the deepest recorded point being little more than ten metres, the lake's size doesn't just vary throughout the year, but has alternately shrunk and grown enormously over the centuries. Being an 'endorheic' system, or a closed drainage basin, means the lake retains the water as there is no outflow or discharge into rivers or oceans, so its precise level tends to be dictated by the difference between any inflow and the amount escaping by evaporation. This will in turn be controlled by the prevailing climatic conditions. The lake is also situated in a region called the Sahel, a climatic zone of semi-arid habitat which runs in a narrow horizontal band across the entire African continent. Representing the transition between the Sahara Desert to the north and the wooded savanna and tropical forests further south, it is likely that this region will represent the first proper opportunity for the Cuckoos to feed since their departure from southern Europe. Using this region to slowly recuperate and recover condition after crossing the world's greatest desert, the satellite telemetry indicates that the Cuckoos will be in no great hurry to reach the Congo Basin with many staying put until around the time that autumn begins to take hold in Britain.

The plumage of the now fully-fledged Cuckoos still back in Britain can vary enormously between being predominantly grey, to decidedly rufous, with the latter coloration not dissimilar to the rare rufous or hepatic colour form of a select few adult females. These juvenile body feathers will not last long though, with many being slowly replaced soon after leaving their hosts' nests. The exception to this rule will be the the wing and tail feathers, which like the adults' plumage will not be replaced until much later in the year. While their errant biological parents rest and recuperate in the Sahel, very little is known about the movements of the juvenile Cuckoos, but it is suspected they may well stay in Britain until at least September. Until then, the young Cuckoos will continue to feed themselves on a diverse array of invertebrate food, with large hairy caterpillars probably the preferred food source. As the summer rolls on, a different suite of butterfly and moth caterpillars will in turn become available for the juvenile Cuckoos to exploit, as they hop around the ground, amongst hedges and up in trees for surely one of the most unpalatable of all meals.

At Puffin colonies around the British Isles, those birds which successfully managed to negotiate their young through to fledging will already have deserted the breeding grounds in their droves by early August, with an exodus rather than a gradual withdrawal being the chosen method of departure at most sites. This will give many well-known puffinries a slightly worn-out and unloved appearance as they become deprived of these colourful and characterful birds for another year. In most seasons breeding success will be dictated by food availability, with between 60 and 80% of pairs managing to see their Puffling successfully fledge in those years with abundant Sandeel stocks. The predation level of chicks tends to be highest at those sites with smaller, less densely packed colonies of Puffins and where the safety in numbers strategy is less effective. Additionally, the synchronised emergence of hordes of young at the larger, crowded puffinries will be more effective at swamping the relatively few predators hoping to make a meal out of a defenceless fledgling.

As the Nightingales, Puffins and Cuckoos desert our shores, the double- and even occasionally triple-brooded nature of many Swallows should ensure that many of these industrious hirundines will hang on in Britain for at least another month before they too feel the irresistible pull of the African continent. As the second clutch of Swallow chicks' demand for food increases exponentially and the female is needed less for brooding this will lead to both parents being suddenly press-ganged into foraging from dawn to dusk. It was of course the famous 'parson-naturalist' Gilbert White who wrote in 1789 that 'all the summer long is the Swallow a most instructive pattern of unwearied industry and affection; for, from morning to night, while there is a family to be supported, she spends the whole day in skimming close to the ground'.

Swallows are exceptionally well adapted for a mode of flight that demands plenty of changes of direction, and when foraging this manoeuvrability is achieved by constantly switching between flapping and gliding. While flapping uses more energy it is of course much faster than gliding and so tends to be used as the main technique. This energetic mode will be interspersed with short gliding periods rarely lasting longer than a couple of seconds. When seen skimming across a field chasing insects, Swallows can give the impression of flying quite quickly, but this is deceptive as their speed is usually clocked at little more than a relatively sedate 8 to 11 metres per second. Their speed will of course vary according to the insects they are pursuing, as fast flight will commonly be needed to catch mobile insects such as horseflies, hoverflies and bluebottles, while a combination of slower flight and gliding will be more than enough to hoover up the weaker flying aphids and midges. Most insects are also caught by the Swallows flying upwards to take them from below, and this is thought to both prevent the insect diving as a means of escape and also to help the bird pick up their quarry more easily against the sky rather than a backdrop of vegetation. Befitting a species that spends a significant part of its life in the air, Swallows have also been calculated as being far more energy efficient than many other birds in flight. Their streamlined shape, relatively long wings and low wing-loading all help reduce the power that the Swallows need to stay airborne. When their long wingspan in relation to the wing area (or high aspect ratio) is also factored in this helps the Swallows to fly slowly without stalling and also when turning to catch insects. These tight turns are also ably assisted by the birds' fanned tail, which when spread, lowered or twisted can give the vital micro-adjustments considered so crucial as they hone in on their target. Finally, their broad bill will help ensure many more strikes than misses.

After close to four weeks in their fortified chamber, many young Kingfishers from second broods will be ready to fledge by early August. Increasingly cramped in their ever more pungent surroundings, the nesting chamber will be barely fit for purpose by the time the young finally decide to take the one-way ticket down the tunnel and into a world of untold danger. Looking not too dissimilar from their parents, there are a few features that do set the juvenile Kingfishers apart. Overall they tend to be duller in colour, with a dirty wash across their upper chest, have much darker feet and a conspicuous white tip to the bill. It will also be some time before the juveniles have mastered the characteristic shrill whistle so distinctive of the adults, and in the intervening period will have to make do with a 'chip' type call which they employ to stay in contact with their parents. Initially they will harass their parents mercilessly to be fed, but it will be only a few days before their parents tire of those antics, forcing the young to either rapidly learn the art of fishing or go hungry. Becoming waterlogged and drowning is a very real threat to the young apprentice Kingfishers as they learn their trade, and they will also need to develop a proficiency in stunning their prey before swallowing in order to ease the fish's passage down their gullet. Another key life skill will be understanding the importance of keeping their plumage in tip-top condition by regular preening. Like the youngsters of any species, some juveniles will learn faster than others, and in the 'dog-eat-dog' world of the Kingfisher it will be the birds that are both keen to learn and take pride in their appearance that will have a much higher probability of surviving through the winter and beyond.

Still very much a family party, the Bewick's Swan cygnets feeding away in the perpetual daylight of the Arctic Russian tundra will be growing quickly with the arrival of the first days of August. The family will be feeding on sedges, grasses and the berries of plants like Cloudberry, Cowberry and Crowberry, which should all be ripening by now. But while foraging they will still need to be constantly alert to the presence of predators across this flat, unforgiving landscape, particularly as both parents may still be flightless. Despite the long days, the temperature will already have begun to drop, with the closest town of Naryan-Mar, for example, falling from an annual high of around 13.5°C in July to an August average of 10.5°C. Dropping quickly after this shortest of summers the temperature on the swans' breeding grounds will then probably remain below 0°C for the entire period between October and April, only thawing once the swans return in May.

Also preferring to keep it in the family on the breeding grounds, the adult Waxwings will only just have begun their annual moult, which is likely to continue through to November and possibly even December. The extended nature of this moult means that in all likelihood most of the Waxwings arriving in Britain, from October onwards, will still be actively replacing their feathers on the move.

Also carrying out a partial moult of their body feathers, most of the young Waxwings will still be quite some distance behind any early fledged British Robins which should just be applying the finishing touches to their moult by early August. Shedding their spotty plumage, as the distinctive orange-red spreads from the upper centre of the breast outwards and downwards, the Robins on completion of their first moult should be pretty similar to how their parents looked in the breeding season. Adopting the badge of maturity ahead of the adults who are still in mid-moult will see the young pretenders looking to steal a march on the 'seniors' as they set about trying to hold a territory for the very first time.

Those Peregrine Falcons that fledged earlier in the summer will still maintain loose bonds with both their parents and their natal site into August, but as they continue to become ever more independent and explore sites further afield, following their progress becomes more difficult. By now, the flocks of moulting Lapwings will also have become highly mobile as they continue to exploit the best feeding areas either close to their breeding grounds or further afield. Certainly British-breeding Lapwings have some of the least understood migrations, as the population will contain a combination of flocks moving within Britain and also birds migrating either to Ireland or even the continent, leading to a very complex picture. However with a plethora of feeding opportunities, plenty of warm days and the majority of continental Lapwings still to arrive, this will inevitably become a time of intense feeding in preparation for the colder weather to come.

By late summer, many juvenile and adult Blue Tits will have merged to form flocks with possibly dozens of other birds from a variety of species. With the food patchily dispersed, a single Blue Tit could spend much of the day attempting to track down the best feeding sites, but its chances of finding good food supplies will be vastly increased by joining forces. Linking up with Great Tits, Coal Tits, Goldcrests, Wrens and even Treecreepers, as well as Chiffchaffs and Blackcaps before they migrate, the mixed flock can work a circuit of woodland, scrub or rural gardens in the full knowledge there will also be more pairs of eyes on the lookout for danger. The different species frequently travel and feed together with little conflict as they will tend to occupy subtly different niches when foraging. Often losing sight of one another as they each search amongst the foliage, twigs, branches and trunks, they will listen out for the contact calls as the flock moves on; being left behind would make the individual more vulnerable to predation. Feeding with acrobatic dexterity on the outer branches and flimsiest of twigs should enable the Blue Tits to hopefully find sufficient food while ensuring they don't either compete with the other species in the foraging flock or become food for Sparrowhawks themselves. The only other species which might compete for the same feeding niche as the Blue Tits, such as Coal Tits and Goldcrests, tend to be either subordinate in the dominance hierarchy, or in the case of the Long-tailed Tits, prefer to keep to their own kind.

Having hatched back in early April, the young Tawny Owls will finally be taking the first tentative steps away from their parents' territory. Initially they may just roost away, only to come back to the site they know best to hunt, but before long they will will bid farewell to their parents and move out permanently. On departure, it's thought the youngsters will explore the wider surrounding area as they scout for a location in which to settle permanently. However, extensive ringing, radio telemetry and satellite tracking has shown the juvenile Tawny Owls will rarely move far from home, with the most common distance between where they hatched and ultimately settled only around 4km. By collating all ringing records, the BTO discovered that a mere 7% of young Tawny Owl recoveries were found more than 20km from where they hatched. The record movement of 687km was by a young Tawny Owl ringed in the Scottish Highlands as a chick, only to be later recovered dead in Dyfed, Wales. The likeliest explanation for this extraordinary distance is that the owl must have been accidentally hit by a vehicle, before then becoming trapped until it finally became dislodged in Wales.

Rarely breeding until at least two years old, the best hope for any dispersing juvenile will be to either chance upon a vacant territory, or replace an adult that hasn't survived the breeding season. For those not lucky enough to immediately strike 'territory' gold, there will always be the fall-back position of becoming a non-territorial 'floating' bird, in the full knowledge that if they get caught by the resident pair they will be driven out with little mercy!

Mid-August

By the time that British school holidays are well under way, the majority of seabird colonies will already have become Puffin-free zones, as the birds abandon their colonies and head off out to sea. Having bid farewell to terra firma for another year, many will have already begun their annual moult, resulting in a far more muted and drab winter plumage. This new coat – and perhaps this is no coincidence – seems to match the dull, uninspiring colours prevalent in both the North Sea and North Atlantic throughout winter. Perhaps the most noticeable change between the Puffin's appearance during the breeding season and throughout winter can be seen on its bill, as the bright orange-red sheaths are sloughed off to reveal a far more slimline black and dull-red beak. In addition, the bright red ornaments around the eye are shed, the cheek rosettes at the base of the bill shrink and fade, the feet lose their bright orange coloration and dusky-coloured feathers emerge from the previously pale grey face patches. As the birds take on a far more sombre appearance, the only feathers retained from the breeding season will be those directly responsible for flight, which will not be shed until much later in the year.

Recent research seems to indicate that British-breeding Puffins will disperse widely upon leaving the colony, but thanks to the BTO's work tracking British-breeding Nightingales with geolocators, it seems they may pass along a far more defined migratory path as they head south through Europe. Data collected from the ground-breaking Nightingale 'OAD' indicated that by the middle of August, having initially arrived in France, it then set off in a direction towards the Iberian Peninsula. Following a similar route to the migratory path Nightingales are believed to take when travelling north in spring, it seems that Spain and Portugal are not just exploited for rest and recuperation after crossing the Sahara Desert and the Mediterranean Sea, but also in preparation for tackling them on the southward journey too. Travelling south at a much more sedate pace compared with the race to get to their breeding grounds in spring, by now the British-breeding birds could well be mingling with a much larger contingent of French Nightingales also bound for Africa. Boasting up to a million pairs of Nightingales each summer, the sheer size of the French population illustrates how Britain is, and probably always has been, right on the edge of the species' natural range.

Having departed Britain well before the Nightingales, it is no surprise that in most years the vast majority of Cuckoos will have already moved south of the Sahara Desert by the middle of this month. For example in the four years between 2011 and 2014 that Chris the Cuckoo was tracked during his southward migration, only in 2014 did he not manage to reach Chad by mid-August. Chris's late arrival into Africa that year also seemed to have mirrored the progress of many of the other tracked Cuckoos that summer, which were also observed delaying their departure from Europe for reasons unknown.

Land-locked by Libya to the north, Sudan to the east, the Central African Republic to the south, Cameroon and Nigeria to the south-west and Niger to the west, land-locked Chad is Africa's fifth largest country. Roughly split into three major geographical zones, the northern third of the country comprises desert in the form of the southern Sahara. This arid region then gives way to the Sahelian belt in Chad's centre, followed by the Eastern Sudanian savanna in the wetter southern third of the country. While the Sahel consists largely of semi-arid steppe dominated by palms and thorny acacia bushes, the higher rainfall in the savanna region creates a far more fertile environment, resulting in a much richer diversity of wildlife. Mostly consisting of large swathes of grassland ideal for grazing, in addition to elephants, rhinoceroses, giraffes, antelopes and many of the large African carnivores, this hugely diverse region has recorded over 500 species of birds and over 1,000 endemic plant species. The Cuckoos are thought to time their arrival in the savanna to coincide with the middle of the region's wet season, which is brought about by the 'intertropical front' operating between May and October. Driving a huge flush of plant growth, the arrival of the rains will also promote an abundance of attendant larvae, such as caterpillars, which theoretically will then be presumably targeted by hungry Cuckoos keen to refuel after such an exhausting flight.

Fledging around the middle of August, the warm temperatures mean there should still be more than enough invertebrate food available for the Swallows' second broods to catch as they come to terms with the joint tasks of both defying gravity and eating on the wing. From extensive data collected by bird ringers, it does seems that these later broods tend to experience much lower survival rates compared to the first broods fledging earlier in the year. Based on the number of birds returning the following spring, a number of theories have been put forward as to why Swallows that fledge later might have drawn the 'short straw'. First, the difference in diet between the first and second broods seems to be an important factor. The food available to Swallows of course will vary throughout the year, with an increase in aphids and a decrease in large flies observed as the summer proceeds. As the high nutritional value of large flies makes them far more profitable to hunt, their decline in abundance throughout the summer may mean that any fledged Swallow will have to work harder for the same nutritional return. This is compounded by the fact that in mid-August, the days, and available light for foraging, will be much shorter than in late June. Also for those second and even third clutches reared in the exact same nest-cup as an earlier brood, an increase in the number of blood-sucking mites could also affect the general level of fitness of the chicks. Any small and subtle difference in the juveniles' health could mean the difference between life and death as the Swallows push their bodies to the limit during the physically stressful and hugely demanding autumn migration.

By the middle of August, the power of flight should finally be returning to the adult Bewick's Swans as they continue replacing their flight feathers up on their Arctic Russian breeding grounds. For those parents with cygnets, this should enable them to once again step up their level of protection against any marauding predators. Now over a month old, the surviving cygnets should be swiftly replacing their off-white down for their first full set of feathers. Sporting a dusky grey plumage and a flesh-coloured bill, the young are easily differentiated from the snow-white plumage and characteristic yellow and black bills of their parents for some considerable time. In fact the full adult plumage will not appear until at least their second winter back in Britain, and with a couple of additional years passing before they will even contemplate a first breeding attempt, the swans could easily be at least six or seven years old before they are sufficiently experienced to see their own young successfully through to fledging.

By now freely moving around the taiga forests, the family parties of Waxwings are now believed to start aggregating into larger flocks as they go about their regular routine of feeding and roosting. This flocking not only serves as protection against predators, but is also thought to operate as an informal information exchange, with hungry birds following those well-fed individuals which have patently located good feeding grounds out of the night roost. Not content with just the company of their own species, the Waxwings may also associate at this time with other species which have bred in the forests of northern Europe, such as the Fieldfare. This noisy, gregarious and garrulous member of the thrush family has always remained an incredibly rare breeding species in Britain, with no more than a handful of confirmed records at a few remote sites in Scotland and northern England. However, each winter up to a million birds will cross the North Sea to help themselves to our berry crop, making them a far more abundant and regular visitor to Britain than the erratic Waxwings.

Becoming ever more adept at feeding themselves, the young Peregrines will be continuing to venture further afield in late summer, safe in the knowledge that the 'comfort blanket' of their parents' territory will still be there if necessary. While some may attempt to breed the following spring, the majority of juveniles are not thought to give it serious consideration until at least their second year and so will join a non-breeding population, whose numbers, distribution and movements are still little known. From studies of the movements of young Peregrines compiled by the BTO, the median distance travelled by over 500 ringed birds was just 45km, with only around a fifth travelling more than 100km from their birthplace. Despite ringing records suggesting that juveniles do not travel too far, this post-breeding dispersal could account for the old falconer's terms of 'passage hawk' for any falcon in immature plumage observed in non-breeding areas. The movement of young British Peregrines over relatively short distances contrasts with other populations of Peregrines found, for example, in North America and across northern Europe and Russia, which are highly migratory, giving clarity to the derivation of the Peregrine's scientific name *Falco peregrinus*, which translates as 'wandering falcon'. These young birds will ultimately hope to fill the gaps caused by mortality in the surrounding breeding populations, and if lucky, skilled and tenacious enough, may even go on to establish new breeding sites themselves.

Also leaving behind all they've ever known as they attempt to either carve out their own territory, or at the very least stay out of trouble, the young Tawny Owls will finally leave their parents behind. Left alone to finish their moult in peace, it is doubtful that the parents will ever venture far from their established territory, as even when prey is thin on the ground, the pair would rather stick than twist. This incredible site fidelity may be one of the reasons why no ringed Tawny Owl has ever been recovered abroad or even foreign-ringed individuals found here, suggesting that the British population of Tawnies is a particularly isolated one. This may indicate that at some point the British Tawny Owl could even become a candidate for being a separate subspecies – with *britannicus* surely a suitable epithet!

Chased and harried by their parents, the local Kingfishers should once again briefly turn their waterfront into a blur of blue and orange as the second brood are chased out by their parents in a none too subtle fashion. The evicted juveniles, however, will have a few options – they can either find their own uninhabited stretch of river or lake, move to the coast where there is less competition or stay and fight for a stretch of river already occupied. Attractive territories, while being rich pickings, are often fiercely defended, and very few end up being wrestled away from older and more experienced birds, but it is not unprecedented. Wildlife cameraman Charlie Hamilton James, for example, observed a very strong-willed subadult female, which he thought came from the year's first brood, that managed the feat of ousting the established resident pair holding territory adjacent to his house in the West Country. Once the young have dispersed, the river may suddenly and temporarily fall very quiet as the adults concentrate on completing their moult in peace, only too aware they will soon have to go back into battle to re-secure their own winter territories.

While fish in the rivers should still be plentiful for any homeless juvenile and moulting adult Kingfishers, this time of year can often present real difficulties for those species which feed principally on soil invertebrates. In a hot, dry summer the soil can quickly turn hard and baked, making soil conditions far more difficult for birds such as Lapwings to forage for prey. Additionally, with many arable fields covered by wheat, barley, oats, rye and rape in summer, the soil also becomes more difficult to access until the crop is harvested. Struggling to find enough food in farmland, feeding flocks during these summer droughts may instead be forced to feed in grassland or wetland habitats, with the result that normally peripheral habitats like reservoirs, gravel pits and sewage farms can often see an influx of hungry Lapwings at this time of year.

Despite the high temperatures in the middle of August, which would theoretically result in an abundance of invertebrate food, summer and autumn can paradoxically be a time of high mortality for inexperienced birds such as juvenile Blue Tits. Deprived of the safety net that their parents provide, many young will run a high risk of either being taken by predators, or dying from starvation or disease brought about by their poor condition. With a staggering 90% of young Blue Tits thought to perish before their first birthday, those managing to survive until November will have already found their odds of reaching the following breeding season to have improved considerably.

Falling silent for just a few short weeks, the Robin's song can once again be heard from the middle of this month, as the younger birds attempt to nail down a winter territory. Distinctive for producing two different types of song, the Robin's so-called 'winter song' which is generally heard from now until late December, tends to be far softer and more introverted than the male's full-blooded 'spring song'. This latter song is a more strident and confident expression of territorial ownership, and is designed to convey the joint message of both intimidating neighbouring males while attracting any nearby females. Despite the 'winter song' undoubtedly being more subtle, the message it conveys is no less important, as it will be the key to ensuring both a meal ticket and even a lifeline right the way through the depths of winter to the following spring.

Late August

Exactly why Waxwings spread across Britain and northern Europe in some years but not others is still little more than educated guesswork amongst many ornithologists. In most years the birds will usually winter close to their breeding grounds in the taiga forests or a touch further south, where temperate forests begin to take over as the dominant biome. However, following a number of good breeding seasons, the Waxwing's population may increase quickly, causing pairs to nest further south than their normal range. The Rowan is believed to be a key species for the Waxwing to feed on during the cold, dark winter months, but it seems that the tree may alternate between good and poor fruiting years. So, when a sustained increase in the Waxwing population combines with a poor fruiting year, such as prior to the mass invasion of 2004/05, then a large number of birds will be forced further south than normal to find sufficient food.

The French ornithologists Jules Fouarge and Philippe Vandevondele carried out a study of the exceptional invasion of Waxwings across Europe during the 2004/05 winter and found that the first tell-tale signs of movement had already been noticed by the end of August. Indications of a ground swell in movement were first observed at sites like Luleå, on the shores of the Gulf of Bothnia in northern Sweden, and at Lake Ladoga, north of St Petersburg in the Russian Federation. The fact that these two watchpoints normally observe Waxwings is not exceptional, but reports of such high numbers passing through even before September seem to provide clear evidence that Britain should prepare for a deluge of these benign invaders.

The immigration of Bewick's Swans into Britain is a far more consistent phenomenon when compared to the irregular arrival of Waxwings, but with cygnets still not yet ready to fledge, any adults with young will still be tied to the Arctic tundra for a few weeks yet. By the end of August the days of perpetual daylight in June will also have disappeared, and as the nights start closing in once more, the current darkness of around nine hours a night will quickly increase with each passing day right up to the winter solstice. As temperatues, and the hours available for foraging, begin falling quickly, it is imperative that the young fledge before the ice returns to the maritime waters of the Barents Sea.

Even for those Blue Tits summering up in the north of Scotland, the breeding season will have long since wound down, giving both adults and young plenty of time to complete their moult before the nights draw in substantially. Having started their slow annual moult possibly even before their young had left the nest, the adults, certainly in southern Britain, should by now be close to sporting a fresh new look. Having not themselves begun their partial moult until late July, it will be at least another six weeks before any southern youngsters begin to resemble their parents in the plumage stakes. Still trying to remain as inconspicuous as possible while they learn the life skills necessary, their dull plumage should not just help them avoid detection from predators keen to take advantage of their inexperience, but also ensure they don't elicit aggressive responses from adults of their own kind. As the new body feathers slowly start to appear, this grubby intermediate phase will give the juveniles a decidedly dishevelled appearance as they're glimpsed dashing to and from garden bird feeders or skulking in the shrubbery.

Having already acquired the all-important badge of adulthood that is the red breast, those juvenile Robins hatching from early broods should by the end of the month already be actively singing and chasing other youngsters in an attempt to secure a territory. They won't have it all their own way for long though, because as soon as the adult Robins complete their own moult in September, the younger birds will suddenly be finding their mettle tested against battle-hardened veterans with previous campaign experience.

Like the Robins, Kingfishers will be desperately keen to procure and maintain a territory throughout the winter months. For those adults that have already chased their young away, giving them free reign once again over their domain, now will be the time to quietly concentrate on their moult, which should have started back in July. Taking over three months to slowly and painstakingly replace their feathers, while crucially remaining both able to fly and dive for dinner, those adults moulting later because of delayed broods may even be forced to suspend their moult for the duration of the winter. In these cases any old feathers that were retained throughout the winter are the first ones to be replaced during their annual moult the following year.

Due to the physically demanding nature of moulting, adult Tawny Owls will not want to be wasting any unnecessary energy re-establishing territories while actively growing and replacing their most important feathers, meaning that most woodlands should be 'hoot-free' zones until at least late September. While the moult of the body plumage might drag on as late as December, the feathers responsible for flight should by now have largely been exchanged for pristine versions, leading to the owls now turning their attention to the replacement of their tail feathers. Also in the middle of their own respective moult, those Lapwing flocks foraging amongst arable crops should begin to see easier feeding opportunities arise as crops begin to be harvested from the fields. The resultant ploughing of the stubble, in preparation for planting, should also, quite literally, unearth a multitude of invertebrates which these keen-eyed birds will be able to exploit as and when opportunities arise.

Having suspended feather replacement until their young were fully fledged, the juveniles' increasing independence should finally allow the adult Peregrines the peace and time to quietly finish their moult by the autumn. In many cases it will also be in the adults' interest to remain on their breeding territory all year round where possible, as any falcons 'deserting' a claim could find it occupied upon their return. Their attachment to sites will of course depend on the availability of food throughout the winter, and certainly for those pairs nesting in towns and cities, the presence of feral pigeons all year round may mean there is no need to leave at all.

With many urban sites also able to offer an abundance of roosting sites, it is not difficult to see why available Peregrine territories in some British towns and cities may be in short supply, if not already at saturation. The fact that fledging Peregrines may be finding it more difficult to find suitable territories was recently revealed by one of the first confirmed cases of co-operative breeding in Britain. As shown on the BBC wildlife programme *Springwatch*, a young male raised in 2011 from a well-known site in the city of Bath was then filmed helping to raise his parents' 2012 brood. While a touch clumsy and haphazard due to his inexperience, he was nevertheless able to beg a share of the food from his parents while also enjoying their protective custody. Helping to rear his own kin might also mean that he could be in a prime position to inherit a top piece of Peregrine real estate should any tragedy befall his parents in the future.

As only a tiny minority of ringed Puffins have ever been recovered away from their traditional nesting sites, until recently very little was known about the travels of these mercurial seabirds after the breeding season. When not tied to their colonies Puffins are rarely seen from land, and so were assumed to disperse widely offshore. Additionally, as no large flocks ever seem to be spotted, it is believed they must also spread out at a low density, possibly in small, scattered groups. In recent years there has been a major advance in our knowledge of their movements through the use of geolocators. While having limitations as a technique, due to any birds with the logging devices attached needing to be re-caught and an inherent geolocation error of ±185km, the work of Mike Harris *et al.* studying the Puffins on the Isle of May has revealed they are indeed very mobile. The data-loggers recovered for this breeding site off Scotland's east coast have revealed that after initially remaining in the north-west North Sea, and reasonably close to where they bred, a movement then takes place. The research work revealed that three-quarters of the birds successfully followed then moved around north Scotland, to spend most of the rest of the year well out into the Atlantic. By comparison, recoveries of ringed birds from populations in western Britain seem to suggest that they may disperse even more widely than those Puffins from the North Sea, with west coast birds moving through the Irish Sea and recorded anywhere from Greenland and Newfoundland to France and the Western Sahara. It has also been suggested that the movement of Puffins from eastern colonies to the Atlantic may be a relatively recent phenomenon, brought about by worsening conditions in their traditional North Sea wintering grounds. With sea temperatures rising in the North Sea, this may have lead to the Puffins' food being driven further north, which in turn will have lead to the birds themselves being forced to follow.

For those Cuckoos, such as Chris, that entered Africa via Italy, the end of August should find the majority of these birds fattening up in and around the Eastern Sudanian savanna in southern Chad. The location of this sub-Saharan stopover appears to differ, however, from those Cuckoos that took the Spanish/Portuguese route into Africa. According to the satellite telemetry data these Iberian birds seem to prefer feeding and resting up at this stage of their journey in northern Nigeria, at least 1,000km further west of their Italian cousins. The feeding area for the more westerly Cuckoos seems to fall into the Western Sudanian savanna, a habitat that, while heavily fragmented due to the large human population in Nigeria, is considered broadly similar to the savanna habitat currently being used by the British-Italian Cuckoos.

Over 4,500km away, back in Britain, the adult Cuckoos' progeny should still be gorging themselves on a variety of caterpillars and other invertebrates as they too prepare to depart. A few weeks after gaining independence from their weary host parents, there does appear to be an element of dispersal from where they hatched, as the young Cuckoos spread out in all directions. With a couple of records of

British-reared juveniles observed as far away as Denmark and Germany during this period, this movement is believed to be different from the proper migration, which observations seem to confirm does not occur until early September.

If the movement of Nightingale OAD is representative of the migration of many British Nightingales currently on their way to Africa, then the end of the month should find them in transit towards the Iberian Peninsula. The data from the geolocator of Nightingale OAD certainly indicate that the slow progress of this bird seems to be because it will have been feeding en route, and with little of the haste that proved such a prominent feature of other tagged Nightingales followed to the breeding grounds in spring. Having taken a detour around Paris in the middle of the month, the fourth

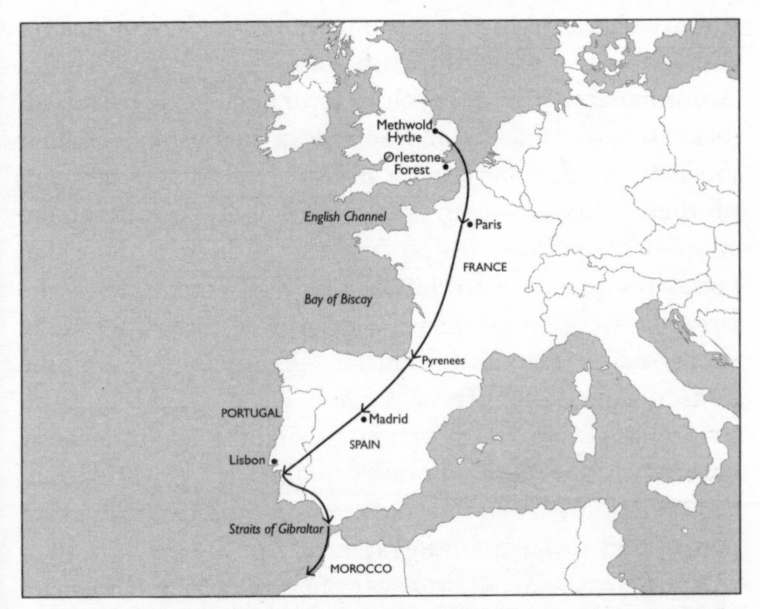

The believed migration route of Nightingale OAD between England and north Africa in the autumn of 2009.

week of August saw OAD skirt past the western end of the Pyrenees, heading down to Madrid before then taking a right-hand turn towards its third European capital city in the space of just a couple of weeks – that of Lisbon in Portugal.

In most years the fledging of their second brood will mark the end of the breeding season for many British Swallows. Finally freed from the constraints of chick-rearing means the adults may keep this second family together for a longer period than would have been the case for their first brood, with some kin possibly remaining together until at least the point when migration gets under way. Still not terribly competent flyers, initially the newly fledged young will need any help they can get as they come to terms with the very real dangers posed by domestic cats, Sparrowhawks and moving traffic. However, after a few days of flying lessons the young should have become sufficiently proficient to spread their wings further afield as they begin a process of familiarising themselves with both the general area and the local Swallow population. Gaining a knowledge of their immediate neighbourhood is not only believed to aid with successfully navigating their return the following spring, but also to help pinpoint any potential breeding sites or roosting spots in future seasons. Certainly for Scottish Swallows, however, these 'fact-finding' trips will not last long, as the deteriorating weather and shorter days combine to push the birds further south. Travelling initially short distances, as they migrate by day and roost by night, it won't be long before the Scottish Swallows begin joining forces with their English and Welsh cousins as the entire summer population prepares for a mass autumn exodus.

September

While the last wave of insects emerge and migrant moths and butterflies take their final fix of nectar before the frosts set in, the first official month of autumn will bring a feeling of change as the slow descent to winter begins in earnest. Bursting forth with nectar and foliage in spring and summer, our trees and shrubs will suddenly become laden with seductive seeds and fleshy fruits. This annual bonanza will have come just at the right time for dormice preparing for hibernation, squirrels stashing for winter and our resident birds looking to lay down fat for the lean times ahead, but too late for the vast majority of our summer visitors, which should by now be busily crossing continents. It's not all one-way traffic though as the relatively mild climate and abundant food offered here will draw a whole new cast of characters keen to spend the winter in Britain.

Early September

Despite a steady stream of continental Lapwings arriving here from as early as May or June, it is not until September that the proper autumn passage of British Lapwings to their winter quarters will begin in any numbers. Continuing right the way through to winter, the movement of Lapwings that breed here is considered to be complex, but by analysing the data from ringing recoveries, Graham Appleton from the BTO seems to have uncovered a few general patterns. Some Lapwings, it seems, are only partially migratory and where possible will winter close to their breeding grounds, while other birds patently favour moving much further in preparation for the coldest months of the year. Furthermore,

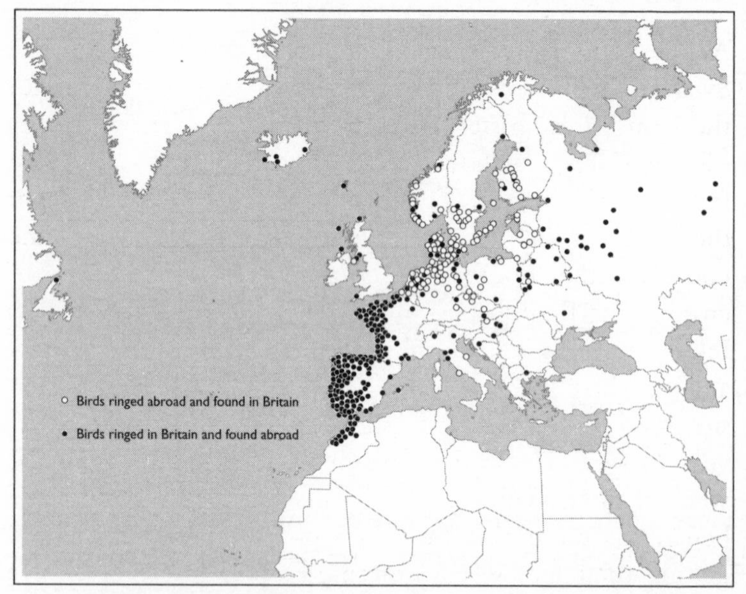

Records of both British and Irish-ringed Lapwings recovered abroad and also foreign-ringed Lapwings subsequently found in Britain and Ireland.

amongst those birds that do prefer to migrate, where they decide to spend the winter seems at least partly to depend on the location where they originally bred.

Many of the Lapwings breeding in the south-east of England, for example, tend to move south to principally coastal locations in France, where the warming effect of the Gulf Steam makes frosts a rare phenomenon, and also to a lesser extent the Mediterranean climate offered down on the Iberian Peninsula. While many of the breeding Lapwings from south-west Britain will also spend the winter on the near continent, a larger proportion of this population seems to prefer crossing the Irish Sea instead to spend the winter in Ireland. For those Lapwings breeding in northern Britain, however, an even larger majority seem to prefer Ireland's winter sun, with few birds believed to prefer crossing to the near continent.

As the Lapwings relocate to their wintering grounds, the Swallows will also begin leaving their breeding sites, with those birds from eastern Britain thought to travel down the eastern side, as opposed to those breeding in the west following the western coastline. Migrating by day, at night the Swallows will coalesce at a number of roosts across the country as they opt for a safety in numbers strategy prior to crossing the English Channel. Some of these roosts, such as Icklesham in Sussex and Slapton in Devon, are well known and regularly used each autumn, but other smaller inland sites tend to be transitory and can often shift location from year to year. The roosting habitats of choice are usually wetland sites, with reedbeds offering the Swallows the benefit of protection from any predators unhappy about getting their feet wet. Gathering at their roost of choice before sunset, the birds initially seem to fly around aimlessly, only changing their behaviour when dusk is rapidly approaching. As the light level drops below a critical level, the Swallows suddenly become galvanised into tighter flocks, which then wheel over the reedbed before small groups can be seen peeling off into the vegetation below. Incessantly twittering to each other in flight, it is only when the birds have settled down for the night that a hush finally descends over the reedbed. Quickly leaving the roost in a series of waves at sunrise, the Swallows will then disperse into the surrounding countryside to take advantage of the year's last flush of insects. These southern roosts may well play host to Swallows for anywhere from a few nights to a couple of weeks, until the shortening days and colder temperatures finally force them to delay their migration no longer.

Keen themselves to escape the increasingly inclement weather in the Arctic, those Bewick's Swans without young still to fledge should be leaving the tundra by early September. The first leg of their journey will often consist of little more than a hop, as the swans slowly gather on traditional pre-migratory staging sites, such as the shallow coastal waters, large lakes and river estuaries close to their breeding grounds. Two of the most important mustering sites are thought to be the Pechora Delta and Korovinskaya Bay in the Nenets Autonomous Region of the Russian Federation, with combined counts of up to 15,000 swans in some years. Both of these sites are located adjacent to the Pechora Sea, and crucially should still be ice-free at this time of year, offering the birds a last safe feeding opportunity before worsening conditions force them in a westerly direction along the coast.

In those winters when large numbers of Waxwings reach Britain, an early sign will be the sheer number of birds seen moving into northern Sweden, from Finland and the adjacent Republic of Karelia in the Russian Federation as early as September. Presumably this extraordinary influx of birds must soon strip out the entire berry supply, forcing a continued push west and south to find as yet untapped sources.

As autumn arrives in northern Europe and the Waxwings are driven on by their stomachs, it should be a very different setting for the adult Cuckoos currently still enjoying Chadian or Nigerian hospitality. In three out of the four years that Chris the Cuckoo was tracked by the BTO, he revealed remarkable fidelity to the wooded savannas of southern Chad at this time of the year, and also relatively little inclination to move around much after his arrival. With the feeding here patently good, the data collected from the transmitters of a number of the birds have shown that most of the Cuckoos are content to stay settled here for at least a few weeks before finally moving towards the Congo Basin. Back in Britain and having jettisoned their foster parents way back in July, the juvenile Cuckoos may well be busily fuelling before preparing to depart on their long and lonely migration to Africa.

Unlike the Cuckoo chicks, the Nightingale young will probably migrate at the same time and along a similar route to their parents, but having severed the family ties back in Britain, the assumption is that it will very much be a case of each Nightingale to their own. From the limited data gleaned by geolocators, it seems that most British Nightingales will spend early September in the very same dry, uncultivated land that the adults themselves will have passed through during their spring migration earlier in the year. Keeping a low profile as they hop around the mosaic of dry scrub and wooded Mediterranean groves, their simple mission will be to put on as much weight as possible before departing from Europe for Africa at the end of the month.

With the Puffins continuing to disperse far and wide from their breeding grounds, those young Peregrines moving away from their parents' territory will now be joining a floating population of birds with a nomadic existence. Being both too inexperienced to hold a territory and still not ready to breed for at least the first couple of years, their only aims will be to find enough food to eat and a safe place to roost without ruffling the feathers of any incumbent Peregrines they may meet on their travels. Peregrine expert Ed Drewitt reported, for example, that at least five different Peregrines were identified using one particular church in the town of Aylesbury, Buckinghamshire during the autumn of 2010, a number of which will certainly have been itinerant birds merely passing through. Due to the difficulties of separating individuals purely on plumage, this will probably have been an underestimate of the real number of Peregrines visiting the town, as some may have been present for little more than a few hours or just a day before then moving on.

Any young Tawny Owls that left their parents' territory earlier in the summer will also be in a similar position to those immature Peregrine Falcons, in that finding a place to feed and safely roost while not alerting any established territory-holding pairs as to their presence will be of the highest priority. With the odds seemingly stacked against the young Tawnies surviving at this stage they have a couple of factors working in their favour. First, autumn is considered to be the peak period of abundance for many of the Tawny Owls' main prey items – mice and voles. The Wood Mouse, for example, is capable of producing litters of four to seven young in successive pregnancies from March to October, leading mammal expert Stephen Harris to estimate that a pre-breeding population of around 38,000,000 could easily rise to about 114,000,000 by autumn. This super-abundance of prey should therefore make it straightforward for even the most inexperienced of young Tawny Owls to catch enough to sustain life while honing their technique. Second, as the adults are still actively moulting in early September, they will not feel the need to put as much energy into fiercely guarding their territories as they would later in the year, meaning any young trespassing Tawnies could temporarily be given more latitude than might normally be expected from such a territorial bird.

Crucial to any Kingfisher's survival through to the following breeding season will be the maintenance of a territory in which the sitting tenant has exclusive fishing rights. Now should be the time when many of these territories are actively being secured, with many males taking possession of exactly the same stretch of river they will have held with their mate during the summer. Unceremoniously booted out, the frequently socially subordinate females have little choice other than to hold a piece of river-front that abuts their old territory, or if this is not possible, a stretch of waterside property reasonably close by. In the world of Kingfisher politics, however, adult females should still be above juveniles in the pecking order, with any youngsters making it through to September forced to take whatever half-decent habitat they can hold without being chased away.

The famous ornithologist and Robin expert David Lack, in his wonderful monograph *The Life of the Robin* wrote: 'From September onwards till May, the woodlands, parks, gardens and hedgerows of England are parcelled out into a great series of small-holdings, each owned by an individual or by a pair of robins.' Finishing their moult before the adults, those Robins that hatched earlier in the year will by now be sporting their first red breast and so should already have been practising the fine art of territory acquisition for a few weeks. However, by early September they will suddenly find the competition increases once more as freshly moulted adults also enter the fray. Recently voted the nation's favourite bird, maybe not as many people would have voted for the Robin if they had been made more aware of this bird's huge propensity for violence!

When not feeding or sleeping, Lack believed a large proportion of a Robin's life is taken up with either singing or fighting. After an uncertain couple of weeks while becoming accustomed to the plumage which puts them at least visually on a level playing field with the adults, the juveniles will have suddenly begun limiting both their singing and aggression to one particular area from which all other Robins will be excluded. To combat the adults' experience, the juveniles will often resort to violence much more readily, and although physical, any ensuing fight will rarely result in the Robins being seriously hurt. This fighting is not just confined to the males either, with some of the females keen to acquire a territory being equally pugnacious in keeping all other Robins at bay. Less inclined to fight than those youngsters busily proving themselves, the adult Robins more commonly rely on singing and posturing, with their red breast prominent, as the main means of securing and holding a territory. As individual battles become settled, the countryside will slowly but surely be carved up into a whole series of interlocking Robin dominions.

Confining their territorial behaviour to just the period around when their chicks were reared means there should be little need for any aggression amongst the hordes of Blue Tits roaming local woodlands and gardens in their mixed feeding flocks at this time of year. Having started their moult as early as June, the adult Blue Tits in the southern half of Britain should by now be completing their moult, which will see the males looking particularly dapper in their electric blue caps. Retaining the vast majority of their flight feathers, the juvenile Blue Tits will still not begin to resemble their parents until their body moult is completed later in the autumn.

Mid-September

As autumn unfolds Robin activity will move centre stage, as the experienced adults enter the fray to compete with those birds barely four months old for that all-important territory which could mean the difference between life and death. For beauty and complexity the Robin's song is often considered to play second fiddle to that of the Nightingale's, but while the latter's song is confined to just south-east England and only heard for a brief period between mid-April and the first days of June, the Robin not only geographically serenades most of Britain, but is also heard for most of the calendar year too. When attempting to describe the Robin's song it has been invariably described as 'a distinctive, cheerful and varied warble', but scientists studying the song are discovering its structure has a surprising level of complexity.

When singing, the Robin utters its song in a series of short bursts, with each burst lasting little more than two or three seconds before being interspersed with a short pause. A bout of singing can often last a few minutes and may often consist of dozens of different bursts of song, each of which can be further broken down into around four to six phrases that last, on average, around half a second each. The French ornithologist Jean-Claude Brémond, who has specialised in bio-acoustics, documented the Robins he was studying as having a repertoire of more than 1,300 different phrases, which when placed together in a constantly varying sequence were theoretically able to create an astronomical number of permutations. By experimentally manipulating pre-recorded Robin song, Brémond also discovered that the different phrases tended to alternate between high and low pitch, and despite the birds being potentially able to produce endlessly variable songs, it is highly likely that all the local Robins become familiar with the basic characteristics of the songs of all the birds in their immediate neighbourhood. Recognising a neighbour's voice will mean that a territory-holding Robin need not waste any time and energy on an aggressive response, saving it instead for when a 'new' voice is heard. Any unfamiliar song suddenly heard in a familiar place, which may represent a newcomer threatening to upset the status quo, will then instantly trigger a call to arms for those territory owners within earshot.

In addition to the song's complexity, volume also plays a key part in helping to convey meaning for the Robins at this crucial time of year. Encroaching birds intent on securing a territory will often begin with a cautiously uttered song from low down in the bushes. Only when no opposition is encountered will the newcomer then gradually ratchet up the volume to a level where the song can be belted out from the tree-tops as it attempts to claim the territory for itself. In those territories already occupied, possession frequently seems to be 'nine-tenths of the law', meaning the owner is usually able to drive the interloper from the premises. So clearly marked are these Robin territories that if a bird suddenly disappears, the surrounding neighbours will quickly extend their boundaries to fill the vacant lot, so that the ground will often be fully occupied again within just 24 hours. At this stage the plaintive and wistful 'winter song' is the warble of choice and will be heard right the way through until they change their tune back to the full-blooded spring song around Christmas. Regardless of the seemingly more laid-back tone to the winter song, it is still primarily concerned with conveying one simple message: 'back off… this is my territory'.

Now unshackled from the needs of their youngsters, the adult Tawny Owls will be quietly left to their own devices at this time of year. Sporting fresh flight feathers, the adult Tawnies will need until at least early December before their entire annual moult is finally completed. Remaining on territory for the duration, and needing only to find sufficient food for themselves, it will still be a few weeks before the pair begins reasserting possession of their dominion. Most Tawny Owls are considered to be a model of monogamy from year to year, with a Finnish study revealing that 67% of females only bred with one male during their reproductive lifetimes. It is also likely that a high proportion of the remaining 33% were only forced to find other partners upon the disappearance of their original mate, making it very much a case of 'till death us do part' for many Tawny Owl relationships.

For those established and experienced Tawny Owls, it will be highly unlikely that their territorial boundaries will change much after the breeding season, contrasting with the far more fluid state of Kingfishers' real estate. Antisocial by nature, Kingfishers usually undergo a post-breeding 'divorce', leading to their territories sometimes being split as each individual bird settles for a more modest-sized piece of waterfront throughout the winter months. Unsurprisingly for a bird with a typical lifespan of little more than two years, it probably pays not to put too much investment into a partner that may not even see out the winter. So by using mostly a core area within their domain, each Kingfisher will hope to spend most of the rest of the year quietly concealed amongst bank-side vegetation, only revealing itself either to feed or to chase away any Kingfishers caught trespassing.

With any semblance of territoriality dropped when their young fledged, adult Blue Tits, along with the small percentage of juveniles still alive, will by now be roaming the countryside for food in their mixed species flocks, a practice that will last right the way through until early spring the following year.

By mid-September those juvenile Peregrines that left their parents' territory in the summer will also be continuing to move around the region, as they explore other territories, meet other Peregrines and hone their hunting skills. In Scotland a study of Peregrines by the Lothian and Borders Raptor Study Group showed that females tended to disperse much further, averaging around 80km from where they were reared, as opposed to their male siblings moving only around 48km. Certainly across northern Europe, continental Peregrines are much more migratory than their British counterparts, with many birds both moving to avoid the colder, shorter days and to follow their prey which is also keen to depart for warmer climes. As continental Peregrines relocate, the east coast of Britain can often see an influx of these wandering raptors from around this time, which have opted to take advantage of our relatively benign winter climate. Most of the Peregrines nesting high up in the Arctic are thought to overwinter in Spain or Africa, while those breeding birds from southern Scandinavia are more likely to move to Britain – a classic example of leap-frog migration, where birds nesting at the higher latitudes travel the furthest distance to spend the winter. The BTO has so far compiled records of 53 Peregrine Falcons ringed abroad and recovered in the UK, with 32 of these coming from Norway and Sweden. Just one example of this continental influx was of a Peregrine ringed as a nestling in Arvika, southern Sweden in 2008, which was then recovered injured in Somerset the following autumn.

As British-nesting Lapwings continue to relocate either within the UK and Ireland or simply migrating abroad themselves, the middle of the month should also mark the point when the Bewick's Swan young finally fledge. Leaving the maritime tundra, the families will be keen to relocate to the relative sanctuary of the lakes, estuaries and coastal waters close by. As conditions continue to worsen, the daily average temperature on the tundra may only be 5°C at this time of year, and with each passing 24-hour period receiving seven minutes less daylight, feeding will soon be downright impossible even for birds as hardy as Bewick's Swans. Immediately identifiable by their grey plumage and grey-pink bills, the youngsters will continue to stand out like sore thumbs from their snow-white parents for some time yet, and any pair managing to fledge at least two youngsters will by any definition have had a very successful breeding season.

Due to both the number of predators and precarious nature of the weather on the breeding grounds, any young cygnet able leave the tundra behind will instantly see its chances of survival suddenly take an exponential leap. With their parents continuing to lead by example, the young will have the luxury of personal, on-tap tuition from mum and dad as they set out on the steep learning curve of acquiring first-hand knowledge of their migration route and the good stopover locations on the way to their wintering grounds. Building in number as they coalesce at a variety of key sites around the Pechora Delta, these locations will give the birds a final opportunity to feed before the continually deteriorating weather eventually persuades them to embark for north-west Europe.

Waxwings amassing in northern Sweden will often be one of the first clear indications of an irruption year in the making, but as large flocks continue to strip the finite food resource, their stay there will be little more than a temporary one as wave after wave of birds are pushed further south. Due to the remote nature of these temperate Swedish and Norwegian forests, the number of Waxwings on the move in invasion years is difficult to estimate, with the French ornithologists Fouarge and Vandevondele having estimated an absolute minimum of 10,000 birds to have congregated in Sweden by mid-September before the great British invasion of 2004 and 2005.

Having made mental notes as to the positions of local landmarks and good feeding areas they might need upon their return, many Swallows should by now be leaving in their droves as the declining number of insects suddenly makes Britain a decidedly unwelcome place for insectivorous birds. Heading off across the English Channel, this narrow marine waterbody will represent just the first of many obstacles in an epic journey will take around five or six weeks, and covering in the process close to 10,000km. After bidding the southern English coast farewell, the Swallows will reach the continent in less than an hour, before then pushing along the French coast towards the Bay of Biscay. Stopping off each night as dusk approaches, the Swallows will often be attracted to roosting sites that may already be occupied by birds, and which have been tried and tested by generations of hirundines over the years. Restricting their navigating and foraging to daylight hours, the Swallows will often migrate reasonably close to the ground where their food tends to accumulate in much higher concentrations. As the birds are quite capable of covering anywhere between 100 and 320km a day, it will not take them long to cross France, before then turning east along the northern edge of the Pyrenees en route to the Mediterranean coast. The distance travelled will of course depend hugely on the weather, with the Swallows' progress often being held up by wet weather or unfavourable winds. Alternatively, they may also choose to have a lazy day at particularly rich feeding sites they encounter along the way, such as The Camargue on France's Mediterranean coast, where the feeding may be just be too good to pass on quickly through.

With the Swallows streaming towards northern Spain, the Nightingales, by contrast, will be deserting the Iberian Peninsula in their droves as they make the short trip across the Mediterranean Sea heading for West Africa.

Nightingale OAD, retrospectively tracked by the BTO following the recovery of its geolocator, is thought to have left southern Portugal on 19 September 2009, only then to arrive in the small disputed territory of Western Sahara by the following day, a scarcely believable distance covered of around 1,300km. This movement of OAD in mid-September also tallies with ringing data of Nightingales collected from the Rock of Gibraltar, the British overseas territory on Spain's southern coast, which reports a spike in Nightingales observed at this time of year. The route taken to the Western Sahara can only be surmised, but if the Nightingales were to follow the Moroccan coastline, after crossing the Strait of Gibraltar, this will mean they are able to circumvent both the Atlas Mountains and the Sahara Desert in one deft sweep.

Having conquered the Sahara Desert as long ago as late July, most British-breeding adult Cuckoos will still currently be feeding up in the wooded savannas of Chad and Nigeria as their biological young back in Britain will finally be setting off for Africa. With no lead to follow, the juveniles' journey must be genetically hardwired, but as satellite technology has still not been miniaturised sufficiently for immature Cuckoos to be tracked, we can only presume that a similar route to that of their parents must be taken.

So little is also known about the dispersal of Puffins from their breeding grounds that their winter distribution must also be considered little more than educated guesswork. However, the limited research carried out suggests that those dispersing furthest are younger birds. Of the four Scottish-ringed Puffins recovered off eastern Canada during the last 50 years, for example, all were in their first year of life. Why young birds seem to have more wanderlust than their less adventurous parents can possibly be explained by the fact that most Puffins do not even attempt to breed until at least five years old. So these far-flung individuals may simply be youngsters keen to see some of the world and enjoy a relatively care-free adolescence until responsibility kicks in a few years down the line.

Late September

Having joined forces to form mixed species flocks with Great Tits, Goldcrests, Wrens and even Treecreepers, back in high summer, juvenile and adult Blue Tits will be keen to maximise the advantage that groups can offer, foraging right the way through the darkest and coldest days ahead. Containing a core number of resident Blue Tits familiar with the area, these roving, mixed species flocks will also include nomadic individuals from further afield, which may come and go as they see fit. Despite their mobile nature, these flocks will nevertheless remain within a comparatively small home range, with almost 90% of ringed Blue Tits recovered being found within 20km of where they were originally caught.

Unlike the roaming Blue Tits, adult Kingfishers will remain rigidly attached to their territory throughout the winter, only possibly accepting any other kingfishers' presence on their patch during the rare occasions when the weather becomes so severe that it will concentrate feeding to a very few locations. After parting ways with their curmudgeonly parents, the importance to juveniles of acquiring a territory cannot be overstated. Most juveniles are not thought to travel far, with few recorded moving more than 12km from their original nest site. During this dispersal period, young birds can occasionally be seen crossing over land as they get ushered out of territories already jealously guarded, or in the process of branching out suddenly turn up at surprising locations, such as garden ponds.

By late September, Robin territories will be becoming quickly delineated as the birds battle it out, initially with birdsong and then with aggression if newly acquired boundaries are not being respected. One of the many astonishing facts about Robins is that at least half of all the females are thought to regularly sing. This is a feature that is decidedly uncommon in the bird world; the only other British species where females are known to commonly join in the dawn chorus are Starling and Dipper. The fact that female Robins sing was already known before being definitively stated by Charles Darwin in 1871, but this unusual behaviour still didn't change the great man's opinion that bird song was primarily 'for charming the female'. It was not until the 1920s that the naturalist William Henry Hudson argued that the reason why females sing as well as males is that both must hold separate territories in autumn and winter. Even though the singing females tend not to be either as vocal or possess quite the repertoire of their male counterparts, we now realise that Mr Hudson's assertion was indeed the correct one.

By now, those juvenile Peregrines that began moving away from their parents' territory in high summer should certainly have become much more proficient at hunting. Such is the immense pressure for young Peregrines to perfect the art of catching prey quickly that raptor expert Ed Drewitt believes that first-year mortality of juveniles could well be similar to that of relatively short-lived birds like Robins and Blackbirds. However, for those young that do manage to make it through the perilous bottleneck of their first year, a longer life expectancy can then be expected, with the current longevity record being that of a male Peregrine ringed as a nestling in Cumbria, which reached at least 17 years and two months.

The most commonly recorded technique for a Peregrine catching and killing prey is with their mightily impressive stoop dive. This method involves the falcon closing its wings to form an aerodynamic teardrop in order to descend at huge speeds as it aims to catch its prey by surprise. Recent research work suggests that Peregrines have learnt to attack in a spiral, which enables them to keep a constant eye on their prey, without frequently turning their head, which could increase drag to such an extent that the inevitable slowing up might result in their quarry potentially escaping. Even though this curved flight path is slightly longer than the straight line, the extra speed generated by significantly reducing the air resistance will more than compensate for the extra distance as they attempt to strike lucky. The Peregrines additionally possess a nictitating membrane, which helps keep their huge, sensitive eyes both dirt-free and moist during the descent. Unlike an eyelid, the translucent nature of the membrane will still enable the bird to detect light while the eye is covered, meaning their eyes can be cleaned without a temporary loss of sight at the crucial moment. The Peregrine's aim is to strike the target so hard with its talons that the prey will instantly be killed, but should this not be the case, the hapless victim will usually be quickly dispatched with a bite to the neck, severing the spinal cord. This lethal bite is executed with a sharp triangular-shaped ridge on the outer edge of the upper mandible called the tomial tooth.

How fast the Peregrines travel during these spectacular dives is difficult to measure, but with speeds estimated to top 200mph, this should enable them to easily catch up with any unsuspecting prey, while also comfortably holding the record for the world's fastest bird. Moving at such phenomenal speeds means the birds will also have to cope with incredibly high G-forces when pulling out of any dives or altering their flight path. In fact, researchers have calculated that the birds may experience forces as high as 28G during these manoeuvres, making light of the maximum G-force recorded on a roller coaster of 6.3G, or even the 9 to 12G fighter pilots will experience when executing turns at high speeds. The ways in which Peregrines are able to cope with this immense force without losing consciousness include the maintenance of a very high heart rate during the stoop and also the use of valves in the veins, which prevent the blood draining away from key areas like the brain at the crucial moment.

For young Tawny Owls, mastering the art of hunting doesn't quite require them to subject their bodies to the extreme forces experienced by Peregrines, but their technique certainly requires no less skill. Unlike Barn and Short-eared Owls, which mostly hunt on the wing, Tawny Owls are the classic 'perch and wait' predators. Using a suitable branch as a base, the owls employ a combination of eyes and ears to constantly scan the ground below. Once movement has been picked up, they will then simply drop down on to the prey with the talons doing the damage. Thrust out in front, the talons are fully opened just prior to impact in order to maximise the 'killing zone', and on the few occasions when the momentum of the strike doesn't kill the mouse or vole outright, the powerful follow-up clench of the talons will. The rough underside of their toes also combines with the incredibly strong grip to ensure that any ensnared prey not immediately killed by the strike will have no chance of escaping. Obviously for rural Tawnies which concentrate principally on mammal prey, in those years when mice and vole populations are at a cyclical high then a far larger proportion of young owls will survive. However, irrespective of the abundance of prey each year, a Finnish study of Tawny Owl mortality still showed the odds being stacked against young birds, with just 33% surviving to their first birthday. Those birds which do manage to survive their first year will then see their survival rates leap to 64% for every subsequent year. These enhanced survival statistics can easily be attributed to the greater experience of older birds and the advantage of an exclusively maintained territory.

While the Lapwings continue to arrive in their winter quarters, the Bewick's Swans in an average year should by now be leaving any pre-migratory sites in and around the Pechora Delta of Arctic Russia as they head for north-west Europe. The swans are believed to take a relatively narrow migration flyway as they follow the north Russian coastline west to the White Sea, before then heading south-west across Karelia, via Lakes Onega and Ladoga, and on to the Baltic coast. Satellite tracking of the swans' migration by Dutch scientists in 1998 and by the WWT in 2003 has also revealed that the autumn passage occurs much more rapidly than the spring migration, with Bewick's commonly only taking one proper rest. For example, five birds carrying transmitters in 1998 were recorded flying virtually non-stop to the Baltic countries, a distance of anywhere between 1,800 and 2,300km, with only brief stops along the way, probably to drink. Three other swans tracked during the same migration also stopped for just a few days at the White Sea, making it probably a far less important stopover site in autumn than spring, with two of these birds then also briefly recorded resting up in the Gulf of Finland, just west of St Petersburg and off Estonia's northern coast. Three other swans tracked by the WWT in 2003 were also recorded to fly directly from the Pechora Delta to the Baltic region, with the journey taking just two days. Possible reasons why this migration may occur more rapidly than their return journey include the birds being held up by icy conditions further north in spring. The swans also need more time to feed intensively in preparation for the rigours of the breeding season ahead. While the number of swans seen in Estonia during the autumn passage is much lower than recorded in spring, one site that seems particularly important at this time of year is Lake Peipus, a huge freshwater lake straddling the Estonian and Russian border. Covering over 3,500km, and ranked the fifth largest in Europe, this lake is now considered a remnant of a much larger body of water dating back to at least the last ice age. Averaging a depth of just seven metres ensures the lake quickly warms up but also rapidly cools down, with the result that it can be frozen for over five months of the year. This shallow nature not only produces particularly rich feeding opportunities but also plenty of suitable resting areas for the swans needing somewhere safe to recover from the first and longest leg of their journey. Sightings of ringed birds on the lake suggest that the swans may well spend up to a couple of weeks feeding away on the abundant pondweeds before then finally moving on to either the Netherlands or Britain.

As Waxwing numbers continue to build in northern Europe in an invasion year, resulting in a surge down to southern Sweden, the majority of British-breeding Cuckoos will still be some 5,000km away due south in either Chad or Nigeria. The moult is complex in Cuckoos and still not clearly understood, but while a partial moult of body feathers may occur as early as June and July, most, if not all, of the wing and tail feathers will not be replaced until well after the birds have arrived in Africa. This should mean that at this stage the birds won't just be feeding to recover condition after their migration across the Sahara but also to power the growth of fresh feathers. The habitat that the Cuckoos have plumped for in Chad and Nigeria may well be very similar to that found in Manda National Park, a reserve close to the border with the Central African Republic, and near to where Chris the Cuckoo has stopped to feed on at least two of his southerly migrations. Consisting of woodland-savanna and grassy floodplain, with patchy stands of dense woodland and fringing vegetation, the Cuckoos here may also be rubbing shoulders with the occasional pack of African Wild Dogs and even wild Ostriches which are also resident in this part of Africa.

It was the pioneering individual 'OAD' that first revealed that the Western Sahara, in north-west Africa, may well figure prominently as the first landfall for British Nightingales freshly arrived from southern Europe. Bordered by the Atlantic Ocean to the west, the territory of the Western Sahara consists mostly of desert flatlands, making the Cuckoo habitat in southern Chad and Nigeria seem positively lush in comparison. In fact with no permanent streams, Western Sahara encompasses some of the most arid and inhospitable terrain on the planet. The sparse greenery that is present seems to be largely confined to the Atlantic coastal desert habitat, a narrow strip of land fringing most of Western Sahara's 1,110km coastline which continues south all the way into Mauritania. Despite the climate being extremely hot and arid here, with only low amounts of episodic rainfall, the mists blown in from the Atlantic help maintain a range of succulent shrubs and arid-adapted plants along this 'green corridor', and also permit the growth of lichens both on the bare ground and on the vascular plants themselves. Classified as part of the Sahara Regional Transition Zone, this ecoregion is far more vegetated than most of the Sahara Desert and also forms part of the East Atlantic Flyway migration route, a corridor enabling a whole range of bird species to travel between Europe and Africa.

In contrast to migratory species like the Bewick's Swan, Nightingale and Cuckoo, which tend to travel along a prescribed migration route, the movement of Puffins after the breeding season is considered more of a dispersal, as they radiate out to sea. Until recently it was thought that the Puffin populations from east and west Britain were spatially segregated, both during and after the breeding season. However, thanks to the work of Mike Harris's team with geolocators on Puffins from the Isle of May, off Scotland's east coast, this is now thought not to be the case. Three-quarters of the birds followed from this colony were seen to carry out a major excursion from the North Sea into the Atlantic at the end of the breeding season, leading to the conclusion that there must after all be some mixing of Puffins from eastern and western breeding colonies during the winter. Recent research also indicates that the rich fishing waters around the Faroe Islands may well be an important overwintering location for many British-breeding Puffins. Consisting of 18 major islands and situated approximately halfway between Norway and Iceland, this remote archipelago lies some 320km north-northwest of Great Britain. Despite its diminutive size, the rugged rocky islands and the rich surrounding seas are able to hold over half a million pairs of breeding Puffins each summer. Mixing with their Faroese cousins, British-breeding Puffins will probably be spread out, possibly in small, scattered groups as they divide their time between fishing and roosting.

In contrast to a possibly large number of our breeding Puffins feeding just south of the Arctic Circle, the Swallows which summered in Britain will be streaming towards the equator by late September. Wanting to avoid a long sea crossing, many Swallows will opt for entering Africa from southern Europe via the Strait of Gibraltar. At just 14km in width, this narrow crossing, which also connects the Atlantic Ocean to the Mediterranean Sea, is not just an incredibly important shipping lane with huge historical significance, but is also arguably the most important bird migration route between Africa and Europe. Used by hundreds of thousands of kites, storks and buzzards each spring and autumn, the narrow strait minimises the distance that these large soaring birds need to spend above the sea. Unlike land, water is incapable of producing thermals, so large birds will need to wait for sufficient lift to get them high enough before resorting to the far more energetic technique of actively flapping across the strait until they can once again access the thermals over dry land. While Swallows will have more than enough energy to keep them flapping the whole way, even across the wider sections, the lack of food available for them to forage over the sea may mean that crossing at the narrowest point will dip least into their limited reserves.

Upon reaching North Africa there may then be a brief opportunity for the Swallows to feed in the fertile crescent of land bathed by the Mediterranean Sea before pushing on inland, over the Atlas Mountains and into the Sahara Desert. Taking on 1,500km of the world's greatest desert on a broad front, mostly west of the Greenwich meridian, will once again provide very few opportunities to feed, as they cross the desert interiors of Morocco and Algeria, before then taking on the equally arid countries of Mauritania and Mali further south. Most small migrants are thought to use an intermittent migration strategy which involves resting in the desert during the daytime, migrating at night and then refuelling at any oases encountered along the way. Swallows are, however, thought to continue their day-flying strategy, while taking advantage of any feeding opportunities that may prevail along the way. One such oasis that a number of trans-Saharan migrants will pass through can be found at Ouadâne, situated on the southern edge of the Adrar Plateau. A World Heritage Site, the old town is situated close to an oasis surrounded by Date Palms and Sorghum, which will undoubtedly provide very welcome feeding opportunities for any desperately hungry and thirsty Swallows which may not have fed since leaving North Africa.

October

October is a month of huge change. As the vibrant green hues which dominated much of the countryside all summer rapidly retreat from view, the russets and reds that characterise this most autumnal of months will suddenly catch the eye. For birds, the breeding season will have been reduced to little more than a distant memory, which will differ markedly from some of our best known mammals, as Grey Seals and Red Deer move centre stage in the mating game. This is also a frenetic month for migration, as a huge variety of birds hone in on their winter destinations. Already one step ahead of an array of winter visitors heading to Britain, many of our resident birds, with their moults now completed, will be using this key month to re-establish territories for the colder and shorter days ahead.

Early October

When the seminal *Migration Atlas* was published in 2002, the whereabouts of British Cuckoos in winter was still considered largely a mystery. Up to this date only a single record of a British Cuckoo had ever been recovered south of the Sahara, belonging to a bird ringed as a nestling in Berkshire in 1928 and then subsequently shot in Cameroon in January 1930. It was the ornithological pioneer Reginald Moreau who speculated in 1972 that British-breeding Cuckoos must winter in Africa south of the equator. Recent ground-breaking work by the BTO tracking Cuckoos by satellite has not only confirmed firstly that Moreau was indeed correct, but furthermore transformed our knowledge of the movement of Cuckoos throughout the year. By additionally opening up this information to the public, thousands of interested bird enthusiasts have been able to follow the precise route taken by Chris and his band of brothers both online and in real time.

One of the main revelations when tracking the Cuckoos was the fact that the migrating birds opted for one of two routes, but still with only one final destination. Depending on whether the Cuckoos take the 'Iberian route' or 'Italian route' into Africa will then dictate where they rest up after crossing the Sahara. For those birds taking the more westerly route, Nigeria is seemingly most favoured as a staging point, while Chad was the country of choice for Cuckoos flying into Africa from further east. Despite occupying different countries, both populations of Cuckoos will still have been resting up in locations dominated by savanna. With this entire region's marked wet season running between April and October, the Cuckoos must have fed sufficiently well in order to now be fit enough to undertake the next leg of their mission, as they head off for the Congo. In the four years Chris the Cuckoo was followed by satellite, 2011 could be considered a typical year. After a very settled couple of months in Chad, on 6 October Chris then suddenly upped sticks to fly almost entirely across the Central African Republic, before then pitching down 790km further south, and a distance of just 8km from the border with the country of the Congo. Poised on the edge of the Congo Basin, and just to the north of the main rainforest block, Chris's movements in that year seems to have followed a similar pattern to that of many of the other tracked Cuckoos in most years, whereby on arrival at the portal of Africa's greatest forest, the birds initially seem reluctant to rush in headlong, but instead prefer to proceed at a more leisurely pace.

For those Swallows that managed to successfully tame the Sahara Desert, there will be no resting on their laurels of a job well done, as they continue to push across into the sub-Saharan Sahelian belt, in the full knowledge they still have an awfully long way to go before finally reaching their wintering grounds. Moving through Mauritania and Mali, the precise route the Swallows take is unclear, but with recoveries of 35 British-ringed birds, representing a substantial proportion of all the British Swallows recovered south of the Sahara (but still north of their final destination) it seems that southern Nigeria figures prominently on the route. With most of these records coming from just north of the Gulf of Guinea in southern Nigeria, it would suggest that the Swallows, upon leaving the Sahel behind, will suddenly take a more easterly direction. Travelling quickly through Côte d'Ivoire, Ghana, Togo and Benin, the Swallows will then enter Nigeria as they converge on the region unattractively nicknamed the 'armpit of Africa'. As per their northward journey in spring, the Swallows are most likely to be migrating in small, loose flocks, consisting of a few tens of birds feeding on the move, and are probably only likely to gather in any large numbers at key roosting sites.

By early October most of the adult Swallows will also have begun their annual moult, which due to the birds' need to fly efficiently throughout the entire journey, will be a long and drawn-out process taking potentially up to half the year. Starting with the body and wings, the first feathers to be replaced will come from the back and rump, as their glossy blue-black upperparts suddenly become far duller and flecked with white. Only when halfway through the protracted wing moult will the tail feathers finally begin to be replaced. The timing of the adults' moult also differs slightly to that of the juveniles, which usually wait until they have arrived in their winter quarters before instigating the change that will ultimately see them transformed into adults.

As both the Swallows and Cuckoos make a beeline for the Congo Basin, the Nightingales will have already taken their foot well off the accelerator pedal following their initial leap across the Mediterranean. Believed to be slowly working their way south along the Atlantic coastal desert of West Africa in early October, the Nightingales will be taking advantage of this gentle rate of progress as an opportunity to feed on the way. The few human inhabitants living in this region are likely to be either nomadic herdsmen or fishermen, so that despite much of this coastal vegetation having been heavily overgrazed, the habitat should still be sufficiently intact for the Nightingales to remain true to their skulking character as they forage for invertebrates on the move.

While Estonia and the Gulf of Finland are likely to hold a high proportion of the European Bewick's Swans at this time, other individuals will have travelled even further south to take advantage of a variety of inland and coastal sites in Lithuania, with the Nemunas Delta being considered the most important site. Protected under the Ramsar Convention, this globally important wetland on the Baltic coast is considered by BirdLife to be the most critical bird area in the whole of the country - representing a crucial layover for millions of migratory birds each year. Formed as the River Neman reaches the Baltic Sea, the delta consists of a maze of river branches and canals which criss-cross to form polders and wetlands, in turn creating the perfect habitat for the swans to rest and feed after their long journey. Having completed their wing moult well before they left their breeding grounds, the adults should by now also have replaced their tail feathers. The body plumage, however, will continue to be replaced throughout the entire winter, and in all probability will not be completed until the birds are preparing to leave their wintering grounds in February of the following year.

As the Bewick's Swans recover condition before undertaking the last leg to Britain, the sheer number of Waxwings in southern Scandinavia during an invasion year will cause the entire berry crop to be rapidly stripped. The scarcity of food will then give the hungry birds little option other than to either make the short trip across to Denmark through the Strait of Øresund or make the longer sea crossing over the North Sea to take advantage of the plentiful berry supplies on offer in Britain. With autumn now proceeding at a brisk pace, the Waxwings may not be the only continental European breeding birds being pushed towards Britain.

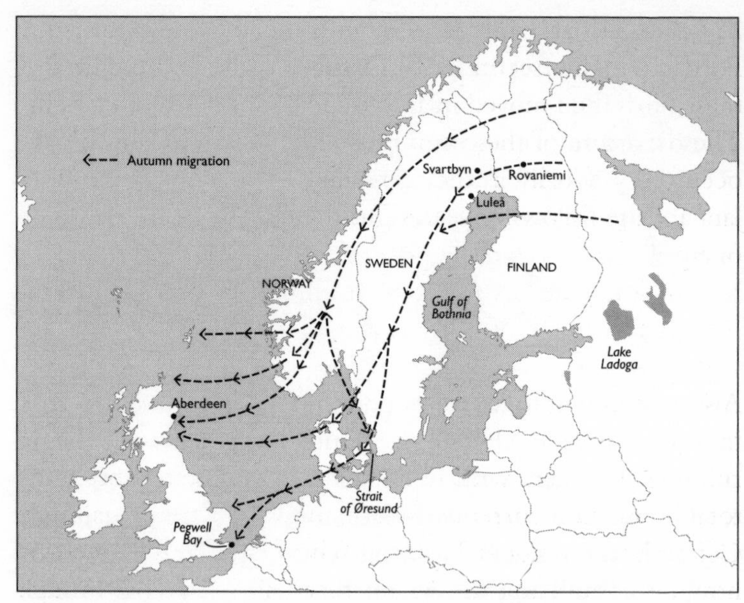

The probable autumnal migration routes of Waxwings between their breeding grounds in northern Europe and their wintering grounds in Britain during an invasion year.

Unlike the largely sedentary British subspecies, migration amongst continental Blue Tits is much more commonplace, particularly in those years when a series of successful breeding seasons can combine to impose such a strain on the amount of food that a substantial number will have little choice other than to inundate coastal areas of Britain. Identifiable with a keen eye, the continental Blue Tits tend to be slightly larger and brighter than their British counterparts and while it's little more than guesswork as to how many of these birds arrive each year, very occasionally the influx can be huge. The celebrated ornithologist Stanley Cramp collated data from the autumn of 1957 when hundreds of Blue Tits were recorded at a whole host of migration watch points along easterly and southerly coasts. With reports of an influx of

447 Blue Tits at Sandwich Bay in Kent and 460 recorded at Portland in Dorset in early October, Cramp thought that these birds then moved inland to spend the winter in Britain. The departure of these continental birds is believed to have occurred gradually, as they returned overseas any time from January the following year, right the way through to March or April.

Also continuing to arrive in good numbers in order to bask in our maritime climate, as continental Lapwings begin rubbing shoulders with resident birds, this should push the total population currently in Britain towards an annual high. Certainly the resident Lapwing which bred here should by now be completing the moult that will see them through winter and beyond, ultimately making both sexes look much more similar until a partial pre-breeding moult early in spring once again differentiates the sexes. The young Lapwings that fledged in the summer, by contrast, will be only halfway through a partial moult not likely to be finished until December, and so can still easily be picked out from the flocks by their shorter crests, scalloped backs and wings and incomplete breast-bands. Like their parents, these immature birds will also undergo a partial breeding moult in the New Year, but will not be finally indistinguishable from their parents until the completion of their first full moult, which will be carried out after their initial breeding attempt.

Needing to complete a full moult themselves, it's likely that the adult Kingfishers will not have replaced all their feathers until November at the earliest. As the old primaries are often not shed until the new ones are sufficiently well grown, the replacement of flight feathers will certainly take far longer than that of the body moult which should already have been completed. When the juvenile moult is carried out will depend on which brood the young Kingfishers were reared from. Confining their first moult to all the feathers apart from those responsible for flight, juveniles from earlier broods may have started growing new feathers by July, while birds fledging from much later broods will probably not be able to start their post-juvenile moult until at least December.

With juvenile Kingfishers leaving the nest during a period spanning more than four months, their moulting process will be a much more staggered affair compared to that of Peregrine youngsters which, irrespective of the location of the nest, are likely to fledge within the month of June. Looking decidedly streaky and brown throughout their first winter, it will not be until they are 18 months to two years of age that the young Peregrines will finally begin to resemble their parents.

As the juveniles make do with their first set of feathers for the time being, the adult Peregrines by contrast should be reaching the final stages of a complete moult that may have begun as early as April or even March. Needing to remain highly effective in the air at all times, the female is thought to begin the all-important wing moult after laying her third egg, while the male will often delay until the young begin hatching. The flight feathers are often replaced carefully in sequence, starting with the fourth of the ten primaries. The replacement then slowly radiates from this point both towards the wing tip and the body. While the body feathers are moulted at the same time as those belonging to the wings, the tail feathers are started later, but should still have been completely replaced by the time the outermost primaries reach their full extent. During this time, the gaps created by feathers being actively replaced will make the adults slightly less efficient at both flying and catching prey. In cases where this severely hampers their ability to catch food, the moult can sometimes be suspended to a time when prey is more plentiful.

As the temperature drops and the nights lengthen, a large part of the Puffins' day is thought to consist of finding enough food to eat as they prepare to change their primaries, an action that will render the Puffins flightless for a short, sharp period, contrasting with the 'slow but sure' technique as exemplified by birds such as Kingfishers and Peregrines.

Having completed their annual moult at least a month before, adult Robins at this stage will still be battling away both with each other and any young Robins for the possession of those all-important territories. With disputes often settled by singing and when necessary fighting, instances of Robins actually killing one another are thought to be very unusual. One of the reasons why mortal combat occurs rarely is that while grappling on the ground they are suddenly far more vulnerable to predation. So it is far more common for the defeated Robin to be ejected with little more than the loss of just a few feathers and a serious dent to its pride as it lives to fight another day.

In addition to the Robins busily jockeying for position, early October will also see adult Tawny Owls exercising their lungs as they begin laying claim to their territory for the winter and beyond. The hoot of a male Tawny Owl is surely one of the most familiar sounds of nature in Britain, as the call commences with a drawn-out '*hooo*', followed by a more subdued '*hu*', which then cues up the final '*huhuhuhooo*' note, which is particularly notable for its strong vibrato quality. This incredibly striking and iconic call can cut right through the silence of a cool moonlit night in autumn, and while the female is capable of uttering a reasonably similar call to her partner, her version tends to be distinctly more strangled and less well phrased. Far more characteristic of the female is the explosive '*kee-wik!*' contact call, which can also be uttered by the male too and is believed to be used as a means of communicating their respective locations in the dark. As well as declaring territorial rights, the various Tawny vocalisations also serve to clearly mark out boundaries, advertise for a mate where the male happens to be single and to further strengthen the bonds of an established pair. Often resorting to duets, this represents a clear message to any owls tempted to trespass, that the territory is not only occupied, but is occupied by a strong, unified pair. When the male does tag his territorial hoot onto his mate's contact call, this is thought to represent the derivation of the '*twit-twoo*' which every schoolchild knows to be the classic sound of an owl!

Mid-October

No two Waxwing invasions ever develop in the same manner, but certainly in most irruption years the first birds will have already made the trip over the North Sea to Britain from southern Norway and Sweden by the middle of October. The first locations to welcome Waxwings tend to be those geographically closest to Scandinavia, with good numbers in Shetland, Orkney and north-east Scotland often providing a clear signal that a massed Scandinavian invasion is imminent. As numbers along Britain's northern and eastern seaboards increase, this gregarious and sociable little bird can suddenly be seen congregating at any sites offering an abundance of food. Led by their stomachs to parks, gardens, supermarket car parks and anywhere berry-laden trees are planted as ornamentals, certainly one place that invariably seems to acquire more than its fair share of these winter visitors is the coastal city of Aberdeen. Nicknamed the 'Waxwing Capital of Britain', a number of Aberdeen's tree-lined streets have become proven 'hotspots' early on in a Waxwing winter.

On arrival in the 'Granite City', having flown either directly from Scandinavia or from earlier staging posts in the Northern Isles, it seems that the Waxwings' first tree of choice is the Rowan. While the Mountain Ash or Rowan is a native species commonly encountered in the woods of northern and western Britain, many of the Rowans we see in our urban streets tend to be either similar species of Japanese or Chinese origin or cultivars of our native species. Commonly planted by councils for their compact shape, and autumn leaf and winter berry colour, these trees provide the perfect welcome mat for this 'pink punk' vanguard as the Waxwings descend to gorge themselves to their hearts' content.

As the first Waxwings take advantage of Aberdeen's street food, the Bewick's Swans should by now be on the final leg of their massive 3,000 to 3,500km long migration. With large numbers having spent the last couple of weeks spread out across Estonia, Lithuania and the Gulf of Finland, a combination of good feeding conditions and lack of disturbance should see the swans sufficiently rejuvenated to head off for their final destination in north-west Europe. Tracking along either the northern, or more commonly the southern shores of the

Baltic Sea, the birds will probably follow a similar migratory path to the one they chose in spring, although the significance of various stopover sites may well vary between the seasons. During the autumn passage, however, fewer records of ringed birds would seem to suggest that the swans do not congregate in the large numbers seen during spring migration, but instead prefer to move quickly on to their final port of call. By observing the departures and arrivals of marked birds, it seems the swans will complete this move from the Baltic staging sites to their wintering grounds in around a week. Many of the swans will fly directly to Britain, but a significant number may also stop briefly to feed on the large stands of pondweeds at sites such as Lake Lauwersmeer in the northern Netherlands. This last pit stop should enable them to top up on fuel before making the short hop across the North Sea to either return to their old stomping grounds, or in the case of the juveniles, become acquainted with Britain for the first time.

Also closing in on their wintering grounds, current research on the satellite tracking of British Cuckoos has shown that certainly most of the male birds (as females are a shade too light to have the current transmitters attached) will have moved into the Congo Basin by the middle of October. Followed for four consecutive years before his disappearance in the Sahara Desert in August 2015, Chris the Cuckoo was not only able to show remarkable site fidelity to his breeding grounds on the Norfolk/Suffolk border, but also a certain faithfulness to wintering in the Western Congolian swamp forests too. One particular area that figured heavily on Chris's itinerary in most winters was around the River Likouala aux Herbes, situated in the north-east of the Republic of Congo. Despite incursions from recent logging practices, this region close to the Congo River is still considered to be heavily forested due to its isolation, making it not only a stronghold for largely undeveloped Pygmy tribes but also a hugely important area for Elephants, Gorillas, Chimpanzees and Dwarf Crocodiles. Still only largely accessible by river ports or the local Impfondu Airport further to the north, the temporarily and permanently flooded swamp forests running alongside the River Likouala aux Herbes contain huge swathes of inundated grasses, giving the adjacent banks the appearance of huge floating prairies. The region is considered to have a year-round tropical climate settling at around 25°C, and with a rainy season running from March to November, the perennially warm and seasonally wet conditions should theoretically enable the Cuckoos to have no problem finding enough invertebrate food to more than power their annual moult.

With the Cuckoos settling down for winter, our Swallows will have no time for loitering to enjoy the delights of the Congolian swamp forests as they press on towards southern Africa. Capable of clocking up to 300km each day, distinct topographical features such as the Congo and Likouala aux Herbes Rivers could be of huge navigational significance as these international jet-setters exchange the Congo Basin's green blanket for the more arid habitats found further south in Angola and Zambia. Well known as a common breeding bird right across Eurasia, hirundine expert Angela Turner reckoned that an estimated European breeding population of between 16 and 36 million pairs could see anywhere from 80 to 190 million Swallows streaming south through Africa during this period. Reliant on flying invertebrates for their staple food, it is thought only the southern hemisphere's warm spring and summer 'austral' temperatures are capable of producing sufficient quantities of food to cater adequately for such a large population of ravenous insectivores, until the lure of the northern spring once again beckons in the New Year.

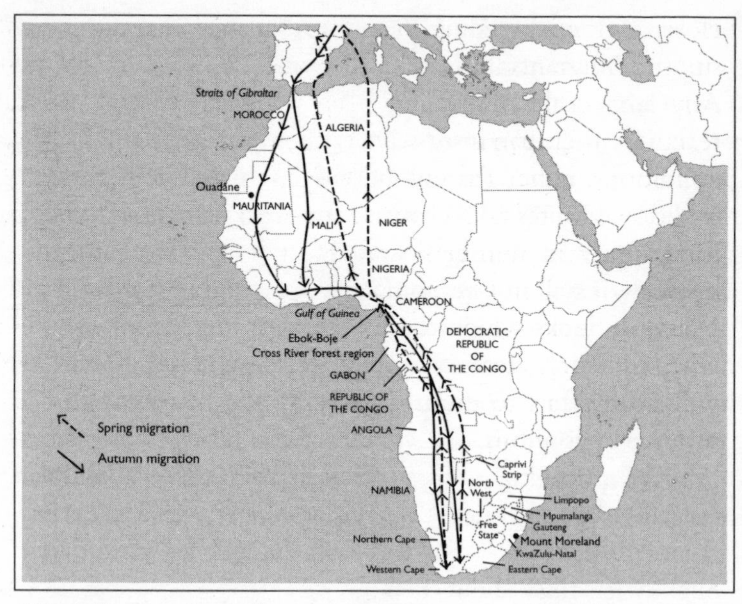

The believed spring and autumn migratory routes of Swallows between their wintering grounds in South Africa and southern Europe.

Having already crossed into Africa at least a month after their huge hop from southern Europe, the mostly ground-feeding technique of Nightingales means that their continued southward journey will need to be carried out at a far more sedate pace than observed in the speedier Swallows. Preferring to 'hug' the coast, these natural-born skulkers may well furtively forage from bush to bush while continuing their steady southward movement through Western Sahara and on towards Mauritania.

Despite a small number of British-bred Lapwings also reaching North Africa each winter, the vast majority of these erigmatic waders will usually choose to stay much closer to their breeding grounds than the more traditional summer migrants like Cuckoos, Swallows and Nightingales. Any large-scale movement of Lapwings also tends to be reactive rather than instinctive, such as in response to freezing conditions, rather than along the far more predetermined, traditional routes taken by most migratory birds. Also opting for a safety in numbers strategy away from the breeding grounds, it will not be until early spring that the gregarious Lapwing flocks which came together in high summer, and which are such a feature of the British countryside in winter, will finally start to disintegrate as the birds return to their summering stations.

Slightly less fussy about the company they keep, our Blue Tits will also be sticking together, as they too remain confined in their mixed species flocks right the way through to early spring. With the adults already sporting fresh plumage after the completion of their annual moult, any young Blue Tits still surviving from the summer's brood should by now also be finishing their partial moult, as they finally resemble their parents in appearance. Not quite a total replacement job, the completed moult will see the main flight feathers and possibly some of the wing coverts still retained from the chicks' original plumage, a handy feature that will enable bird ringers to still identify them in the hand as 'first-year birds'. Only when these immature Blue Tits have completed their first breeding cycle will ringers finally consider them to be inseparable in the hand from fully mature birds.

Unlike the extensive work carried out to understand the moult of common garden birds like Blue Tits, far less is known about the moulting of species like Puffins, due to their disappearance out to sea for the entire winter. As a result of knowledge gleaned mostly from birds either found dead or shot in winter, the Puffin's primary wing feathers are thought to moult in a synchronous fashion, a process which will render the birds flightless for at least four or five weeks. Flightless Puffins have actually been recorded in every month between October and April, with the examination of corpses recovered dead from beaches seemingly indicating a moulting peak in late winter. However, a wreck of 36 dead Puffins from around the Northern Isles after bad weather in October 2007 found 32 of the birds to have been actively in wing moult at the time of their death, suggesting that the majority of wing moults may in fact be more commonly taking place in the autumn. Of course, any Puffin unable to fly will not only be very susceptible to localised weather conditions, but will need to ensure that it has chosen a sea station able to provide sufficient food to keep it sustained during the period when its dispersal powers will have become extremely limited.

Mid-October is of course a peak time for migration in Britain, marking the period when the last of our summer migrants leave, and the first of our winter visitors begin to arrive. Currently close to completing their long, protracted annual moult and with surely tougher times ahead, the adult Peregrines will be keen to cash in on this annual flying food bonanza. In spring, urban Peregrines have been recorded taking summer visitors such as Wheatears, Turtle Doves and even Corncrakes, but in autumn a different suite of migratory birds has recently been revealed in their prey leftovers. As the temperature falls away, birds such as continental Woodcock and Water Rail are now being commonly recorded at a variety of urban Peregrine roosting sites. These normally secretive bird species tend to have relatively short, rounded wings and tails, enabling them to fly quickly but providing poor manoeuvrability, resulting in them having to migrate at night to avoid detection by predators. Because of the immense amount of light pollution now obvious in most conurbations, Peregrine expert Ed Drewitt believes the pale underparts of these migrating birds, which would normally have remained hidden by the cloak of night, are now being lit up like light bulbs as they pass above our illuminated urban landscapes. Perching in the shadows below, many urban Peregrines have cottoned on to this 'nocturnal buffet' and rather than adopting their more conventional 'teardrop' stoop to strike down on their prey from above during the daytime, they have fashioned a novel technique of simply flying up a few tens of metres in the dead of night to snatch the unwary birds from below. During this time of plenty, with abundant and easy kills offered up virtually on a plate, the Peregrines may well now have begun to cache any surplus prey items in the full knowledge that the cooler temperatures should help preserve the food for longer. This natural outdoor refrigerator will then be able to offer up an easy ready meal when catching prey becomes more difficult in the depths of winter.

It will not just be adult Peregrines that are keen to maintain a presence on their breeding territory throughout the winter, as our resident Kingfishers will also be very reticent to budge from any watercourse offering both food and protection as they too prepare for their most challenging part of the year. In locations where most territories have already been earmarked by incumbent birds, the bottom of the Kingfisher pecking order will be those immature birds that fledged in the summer, whose inexperience means they will often be pushed into marginal habitats. Forced to eke out a living in areas away from the dominant adult birds, many juveniles after the breeding season may have little option other than to head for the coast to try their luck. Even though fishing may be more difficult in turbid estuaries or saltwater creeks, the sheer amount of real estate available along Britain's convoluted coastline will often render holding a territory an unnecessary waste of energy. This will then have the upside of leaving the Kingfishers with one less thing to worry about as they concentrate on finding enough food to see each night through. In addition to familiarising themselves with a new habitat, these coastal birds will also have to get used to an entirely new diet as the regular food of Minnows, Sticklebacks and Bullheads become replaced by Gobies, Blennies and Sand Smelt.

As territorial boundaries become clearly delineated and disputes resolved, many Robins by the middle of October will have put their battles behind them and lowered their aggression levels to the less demanding job of territory maintenance rather than acquisition. With some juveniles holding territories alongside seasoned campaigners, there will be little to separate them in appearance, as each bird flaunts its familiar red breast to reiterate its intention to stay put right through to spring and beyond.

Like the Robins, holding a territory throughout the winter will also be vital for the survival of Tawny Owls, but due to the young owls' prolonged adolescence, it could be at least a couple of years before these inexperienced birds are given their first opportunity to get their talons on the first rung of the property ladder. Peaking in late October to early November, the frequency of calling by established pairs will only intensify as autumn gathers momentum, but can vary from night to night depending on the weather. Often less vocal on cloudy and windy nights, the best times to hear the quintessential sound of autumn will be on those warm and still autumnal evenings when the moon is clearly visible.

Late October

Heavily moulting as they migrate, many Swallows will be looking pretty ragged as they stream across Namibia and Botswana's southern borders, en route to their final overwintering destination. With the earliest birds turning up in late August, and stragglers still appearing in December, the majority of Swallows should invariably sweep into their South African quarters towards the end of October. Having fed on the wing for almost the entire length of their journey, the Swallows will suddenly realise that they won't have a monopoly on the 'aerial invertebrate buffet' but will instead have to share the available food with the local competition. Species such as the Greater Striped, White-throated and Pearl-breasted Swallows are intra-African migrants, which having wintered further north will now all be right in the middle of their own respective breeding seasons.

Ringing recoveries suggest that prior to the 1960s, British-breeding Swallows seemed to be overwintering in an area centred around the city of Pretoria, in Gauteng Province situated the northeast of the country, before then making a switch as they spread further southwest. Currently, most of the recent records come from the Western Cape, the fourth largest of South Africa's nine provinces in the southwestern part of the country. Roughly the size of England, the Western Cape is topographically exceptionally diverse and houses an incredibly rich vegetation with one the world's six floral kingdoms almost entirely confined within the province's borders. The Cape Floral Kingdom is often called 'fynbos', a term derived from the Afrikaans for 'fine bush', and refers to the unique vegetation consisting of thousands of evergreen shrubs and flowering plants covering virtually treeless terrain. This incredible landscape is certainly an utterly different habitat to anything the Swallows will have experienced back in Britain, some 9,600km to the north.

Some six weeks after the Swallows departed our shores, and after an equally epic journey back from their own Arctic breeding grounds in the Russian Federation, the Bewick's Swans should also finally be touching back down in Britain towards the end of the month. As possibly two-thirds of all British Bewick's Swans tend to overwinter in either the Ouse or Nene Washes, it is perhaps no surprise that most will enter through East Anglia as they home in on tried and trusted fenland sites that have been used by generations of swans. While far fewer will travel further west, the small population of Bewick's Swans that overwinters each year in and around WWT Slimbridge in Gloucestershire is undoubtedly one of the most intensely studied groups of birds in the world.

By observing and painting these birds from his observatory home at Slimbridge, it was Sir Peter Scott who first realised that the colouring and patterns of each Bewick's bill was as unique to each bird as fingerprints are to humans. Able to place each Swan into one of three basic bill categories – yellow-neb, black-neb and penny-face – within these groups he then looked at infinitely subtle variations, together with a range of other distinctive features, to provide a name for every bird visiting the reserve. It was only when he could systematically identify different individuals that Scott realised many of the Swans were in fact incredibly faithful to

Slimbridge, often returning year after year. Over 50 years after this pioneering research work was initiated, Scott's 'Facebook for Swans' is still being actively compiled at Slimbridge today, with the bill patterns of over 7,600 adults having been recorded and computerised. This ability to identify each individual animal upon its return has also enabled researchers to uncover a wealth of information about these long-lived birds, such as their survival rates, familial relationships and how the hierarchies operate within the flock. The project has shown, for example, that between 40 and 50% of all the adults and yearlings recorded in any one year are birds that will have visited Slimbridge during at least one, or more, previous seasons – in essence Bewick's Swans are creatures of habit.

Upon the swans' arrival there will be two immediate priorities – feeding up to put on weight after the long, physically demanding migration and quickly establishing their position within the dominance hierarchy. Unfettered access to both prime feeding sites and safe roosting locations are incredibly important, but how well the swans eat and sleep will largely depend on their social standing within the flock. With so much at stake upon their return it is perhaps no surprise that aggressive behaviour, in the form of physical and vocal displays, promptly comes to the fore as the swans jockey for position. As they attempt to assert their dominance, any swans still remaining part of a family unit will find that by sticking together they will collectively be able to elevate the status of their group considerably. Those pairs that successfully bred in the summer, and with accompanying youngsters prepared to back their parents in any fight, will use their numbers to form a powerful cartel easily able to bulldoze cygnet-less pairs and singletons out of the way. In fact, so appealing are these family groups that some birds may even roll back the years to re-join forces with their parents on the wintering grounds, in the knowledge that they'll not only be enhancing their own position, but also bolstering their family's standing even further up the pecking order.

Equally gregarious, but with aggression a far less significant part of their daily lives when compared to the quarrelsome Bewick's, the Waxwings will continue to pour across the North Sea as October draws to a close. The city of Aberdeen's location in relation to south-west Norway, and its streets lined with ornamental fruit-bearing trees, means it invariably figures prominently in the early stages of an invasion. Dashing between feeding trees and prominent perches, the Waxwings' fast, direct flight on triangular-shaped wings can make them appear strikingly similar to Starlings when seen in silhouette. Their accompanying 'tinkling bell' contact call, however, should be more than enough to confirm the identity of this most distinctive and delightful of winter visitors, even in poor light.

Unfortunately it is this characteristic low and direct flight that can also make them particularly vulnerable in towns and cities, where a potential hazard lurks around every corner. Waxwings, for example, are considered particularly susceptible when it comes to colliding with windows, and in the invasion year of 2004/05, out of 87 Waxwings reported dead, 66% had collided with these most invisible of obstacles. With multiple fatalities frequently reported from the same site the most perilous locations seem to be when windows are either situated nearby, or directly in between, a regular food source and their favoured perching points, like TV aerials. One particular street in Aberdeen called Wallfield Crescent, where Rowan trees grow between tenement blocks, was nicknamed the 'Waxwing graveyard' after eight were picked up dead over a short space of time during 2008. At another traditional site in the city, a recently erected bus shelter which was unfortunately screened by trees subsequently accounted for a further eight fatalities in the same year, until a Good (ornithological) Samaritan sprayed the back of the shelter blue to prevent any further incidents.

Currently believed to be in the Western Sahara, as they slowly work their way south along the thin coastal strip of vegetation separating the ocean from the desert, flying into man-made objects should be the least of the Nightingales' worries as they traverse this incredibly sparsely populated terrain. Western Sahara is currently a disputed territory between Morocco to the north and the Sahrawi Arab Democratic Republic, strongly backed by Algeria, to the extreme north-east. Consisting of mostly flat desert, the territory contains some of the most arid and inhospitable land on the planet, and despite occasionally experiencing flash flooding in the spring, has no permanent streams. Despite being marginally larger than the UK, the Western Sahara's entire population is less than that of the city of Sheffield, with many living in Laâyoune, the territory's only city of note. Unsurprisingly, given the barren nature of the terrain, employment opportunities are limited, with two-thirds of the entire workforce either engaged in fishing or phosphate mining. Eking out an existence on any invertebrates or berries encountered as they pick their way through the drought-tolerant bushes and shrubs of this politically and climatically inhospitable terrain, the Nightingales will certainly be looking forward to the far richer feeding opportunities offered up by Senegal and The Gambia in November.

Hardly a model of peace itself in recent decades, the Congo Basin should already have been playing host to most of the British-breeding Cuckoos for much of the month. Satellite telemetry has recently revealed that many will spend most of the winter residing in a huge ecological region called the Western Congolian swamp forests, a little-visited habitat centred around the mighty Congo River and which straddles the Republic of Congo to the west and the larger Democratic Republic of Congo (formerly Zaire) to the east. When the Western Congolian swamp forests are combined with similar habitat further east, the combined area is thought to form one of the largest continuous areas of freshwater swamp forest in the world. With these forests considered to be largely pristine due to access being almost impossible, it is likely that the Cuckoos will still be actively moulting as they feed away in these lush, impenetrable forests. At this stage of the year the Cuckoos will have no desire to attract either a mate or attention, so will probably spend most of their time keeping to the shadows and maintaining radio silence as the males save their renowned voices for their return to Britain.

Despite being spread over several countries, the overwintering distribution of Cuckoos is thought to cover a much smaller geographical area than that of the Puffins out of the breeding season, which by now will have dispersed far and wide. Apart from the work carried out by Mike Harris's team placing geolocators on Puffins from the Isle of May, the only other data on winter distribution comes from ringed Puffins that have subsequently been shot out at sea, caught in fish nets or recovered following severe weather or pollution incidents. Eastern Canada frequently reports Puffins from its waters, but with only four confirmed Scottish records in 50 years it would seem the majority of Puffins sighted here are coming from colonies on Iceland, Norway and Greenland. This paucity of records, despite the large numbers of Puffins ringed at some colonies, might well suggest that this huge journey right across the Atlantic may well be more of an exception than the norm. The countries with most recoveries of British-ringed Puffins are France and Spain, with the very limited data suggesting that the majority of these Puffins may well come from south-western locations around Britain, such as the puffinry found on Skomer Island, off the coast of west Wales.

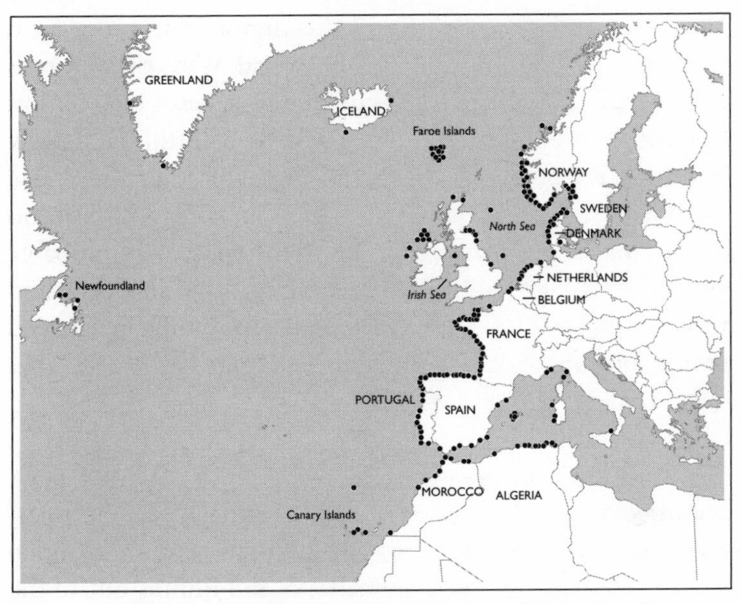

All records of Puffins ringed in Britain and Ireland and subsequently recovered abroad.

Still in their mixed species flocks, the Blue Tits will be continuing to work their regular beat close to where they bred, as they undertake the dual daily challenge of both finding enough to eat, yet avoiding being eaten themselves. Despite the mobile nature of these flocks they're thought to only range over a reasonably small area, with ringing data suggesting that urban and suburban Blue Tits roam even less than their country cousins. This may be down to a variety of factors, such as conurbations being a degree or two warmer than the surrounding countryside, and urban areas offering more roosting spots and feeding opportunities, thanks to the provision of both nest boxes and food by enthusiastic householders.

Certainly it is highly unlikely that urban adult Peregrine Falcons will move far from the site where they raised their brood over the summer either. Once the young have finally dispersed, the reduced hunting burden will leave them free to concentrate on their moult, which in many cases should be approaching completion by the end of the month. Captive birds have been observed taking anywhere between 128 and 185 days to carry out their entire moult, a lengthy period, but one that is deemed essential for any species that must remain airborne throughout the entire process. To ensure the Peregrine will still be able to hunt effectively at this time, the feathers will need to be replaced in a precise and meticulous sequence to make sure any reduction in speed and agility is kept to the absolute minimum. This 'new coat' will not only increase their hunting efficiency, but will also keep them both warm and well protected during the winter months. Aware that well maintained plumage will be one of the keys for survival, the Peregrine will spend a significant portion of its day meticulously preening as it attempts to keep each feather in as good a condition as possible, for as long as possible.

Any established Tawny Owls, which until now will have only been calling sporadically, will shift from first to fifth gear quickly as they reassert ownership of their territory with winter rapidly approaching. The classic hoot is a clear territorial declaration that will quickly be backed up with aggressive intent, and even direct attack, if any intruding bird refuses to beat a hasty retreat. The call's depth of pitch will also vary between individual owls and is thought to be directly related to the bird's body mass. Additionally, the bigger, heavier males are also believed to be able to hold the last tremulous note for longer. Any trespassing owl eavesdropping on these hoots will be able to instantly process all this information in order to decide whether it might either fancy its chances, or simply exercise caution by beating a hasty retreat. Tawny Owls are even thought to be able to recognise individuals on the basis of their calls, enabling each bird to differentiate between a familiar neighbour in an adjacent territory, which may not even elicit a response, and a new owl on the block, which will instantly set off territorial alarm bells.

With most boundary disputes resolved, most resident Robins will by now be fully ensconced in their respective territories for the rest of the winter. Despite a lowering of the state of alert from red to amber, any territory-holding Robin will still be looking to drive out any other Robins caught trespassing, with some landlords busier than others. Interestingly, any Robins holding an inland territory will probably have to issue far fewer eviction notices than those individuals maintaining territories along the east coast, which at this time of year may also have to contend with an influx of continental Robins. The numbers crossing the North Sea will vary enormously each autumn, but in certain years the number of Robins arriving from Scandinavia and the Low Countries can be very high, such as the 1,500 recorded along the coast of Essex over just a couple of days in October 1988. Considered a different race to our resident Robins, the continental birds tend to have a yellowy-orange breast and a far more skulking nature than the confiding British race. Fortunately for our native Robins'

sanity this influx is only usually a temporary one, as the majority of these continental Robins will be doing little more than resting up and foraging before then moving down to spend the winter much further south in Spain, Portugal or even North Africa. The minority of continental Robins that do choose to stick around all winter, however, will pose more of a threat as they attempt to wrestle parcels of real estate off resident birds for the duration of their stay.

Possibly even more antisocial than the Robin, our Kingfishers will currently be spending the entire winter hunkered down alone in their respective territories, hoping that the winter will stay sufficiently mild to prevent their favourite fishing spot from freezing over. Very cold winters can have a huge impact on Kingfisher populations, with sub-zero temperatures forcing birds to either dive through ice or move elsewhere to find food. Even in those winters when the water keeps flowing, the depressed temperatures will make their quarry for more difficult to catch. To conserve energy during the winter, fish tend to be far more torpid and will frequently resort to remaining well hidden on the stream-bed or tucked along the riverbank, making them even more difficult to catch. Needing to eat frequently due to their diminutive size, it has been estimated that Kingfishers will need to consume around 60% of their body weight each day just to ensure they can generate enough heat to see them through the night. With so many demands placed on this tough little bird it is perhaps no surprise that relatively few will live to see their third birthday.

A far more catholic diet will ensure that the variety of feeding locations for Lapwings is way more diverse than for the supremely specialised Kingfishers, but the weather will still play a critical role in deciding the precise habitat this sociable wader is able to use at any one time. The core historic winter distribution for Lapwings was across southern or central England, where mixed farmland containing livestock feeding on improved permanent pasture predominated, but due to the recent run of milder winters, more Lapwings are now being reported principally from arable lands in more easterly areas. The favoured fields of many flocks seem to be those which are cultivated for winter cereals, such as wheat, barley and oats, with oilseed rape rarely touched. While the arable crops seem to have richer feeding opportunities, they are also more prone to frosts, so if the ground temporarily becomes frozen, instantly rendering the soil invertebrates unobtainable, the flock will have a decision to make. Needing to move before risking starvation, the Lapwings will then either take a short flight to a habitat which will be more resistant to frost, such as nearby grassland, or tackle a much longer journey, to a more amenable climate further west. Being able to quickly find and exploit transient food sources provided by cultivation and harvesting means that providing the freeze is not too deep or widespread, the Lapwings will usually be able to find more than enough food to see them through the winter. In fact this uncanny ability to find food will often be taken advantage of by other species, with Golden Plover often using Lapwings to locate the best feeding sites. Large Lapwing flocks will also attract the unwelcome attentions of Black-headed and Common Gulls keen to use and abuse the Lapwings' food-finding attributes before then using their bigger bulk and bullying tactics to steal a free meal.

November

The slow inexorable slide to winter will quickly gather pace in November, as the rapidly declining temperatures and shorter days collude to take a firm grip on the wildlife. Certainly any mammal unable to cope with the cold and lack of food will have long since retired from view to hibernate. The mass-coordinated leaf fall will suddenly see the dreys of squirrels and empty bird nests standing out amongst the bare branches, with Ivy one of the few plants to buck the trend by providing a last splash of nectar for any insects also preparing to overwinter in a deep sleep. Many birds which opted to either stay in Britain or visit for the winter will find that the benefits of flocking together far outweigh those of a winter spent in splendid isolation, and now is the time to see wonderful aggregations of starlings, wagtails, tits, gulls and waders. With the summer migrants long since gone, by November it will be pretty much one-way traffic across the North Sea, as a whole raft of different birds continue to exchange the cold continent for the milder climate on offer around the British Isles, thanks to its warm relationship with the Gulf Stream.

Early November

Having led the charge of the winter visitors for some time, the population of Lapwings in Britain could easily reach over 600,000 by the time November finally arrives. Impressive though this figure might sound, it is, in fact, the lowest it has been for at least a generation and is a sad reflection of an alarming decline in the breeding success of this charming farmland wader not just in Britain, but right across northern Europe too. Lapwings have become one of the most notable victims of the large-scale agricultural intensification in the countryside, as heavily managed crops and high stocking rates of cattle give little opportunity for wildlife to coexist alongside such modern practices.

As both continental and British Lapwings roam the countryside to take advantage of any foraging opportunities, the size of these feeding flocks will invariably depend upon the amount of food available. Within the landscape, these flocks also tend to be highly clumped, with many fields remaining untouched, while others are used intensively. Lapwing expert Michael Shrubb carried out research into Lapwing field-use during three successive winters in Sussex, and found that only 40% of the 111 fields in his study area were ever visited, and just nine fields were used regularly. As good feeding is obviously distributed so patchily, the Lapwing must therefore operate both collectively and effectively in order to track down the best sites, in the full knowledge that when good feeding is found, there should be more than enough to go around.

Upon the flock's arrival at a rich feeding site, it will then be down to each individual Lapwing to forage for itself. Lapwings have large eyes relative to their bill size and tend to hunt primarily by sight. The feeding technique most frequently adopted is known as 'pause/travel feeding' and consists of a few steps, followed by a pause, a scan, and then a move forward to either stoop for a prey item, or to repeat the process if nothing is spotted. Despite this being considered an effective method, not all scans locate prey and not all strikes are successful. When found, the prey will then be prized from the soil by the Lapwing's short bill. In contrast to many of the long-billed waders, such as curlews and godwits, the Lapwing's bill doesn't have a sensitive tip capable of locating food by touch, but is instead hard and horny for grasping and holding the prey. Most of their diet seems to consist of carabid and staphylinid beetles, millipedes, slugs, leatherjackets and earthworms, with the food taken varying seasonally according to the annual life cycles of prey organism. Also the weather conditions will affect the state of the soil, with earthworms, for example, travelling deeper and in the process becoming far more inaccessible in the dry heat of the summer and autumn sun.

Another advantage of flocking, in addition to sourcing patchily distributed food, means that there will be many more pairs of eyes looking out for predators, but one obvious downside to sharing a 'dining table' with so many others means that squabbles between individual birds in too close a proximity will often become inevitable. In denser flocks, aggressive interactions are recorded more regularly, resulting in territorial displays perhaps more commonly associated with breeding grounds also being observed on the Lapwings' winter territories. Two evenly matched birds arguing over a feeding spot will often resort to parallel walking along an invisible boundary as they size each other up. If the dispute isn't quickly settled by the equivalent of this staring match, then it can quickly escalate into hunched or crouched running, jumping up at each other and ultimately fighting.

Continuing to arrive at their traditional overwintering sites such as Slimbridge in Gloucestershire, the Bewick's Swans will also find that a rapidly increasing flock density will initially cause a huge amount of consternation in the ranks until a pecking order finally becomes established. Containing all social groups, the wintering flocks of Bewick's Swans will include families, pairs without cygnets, single adults and yearling birds. When it comes to forming the dominance hierarchy, numbers and experience will count for everything. Disagreements will initially be addressed by ritualised behaviour, but can quickly escalate if neither side is willing to back down. The first sign of a dispute may consist of nothing more than a head lowered in threat, or a well-placed peck, but if these are ignored, then the Swans will enter a well-choreographed and established routine which increases in intensity with each step. The male is invariably the most common protagonist in these aggressive encounters, and after starting with neck stretching and head pumping, he will quickly move through the gears in his threatening repertoire. Starting with extending his wings to a half open position, he will then ramp up the threat to progressively higher levels, to include wings being half open and flapping, wings then fully extended and flapping and finally wings fully extended and still. If by this stage his opponent is still not willing to concede defeat, then the behaviour may quickly descend into a fight with the aggressor biting while beating his wings, in the full knowledge that his family will be noisily lending support in the wings.

As subordinate single swans will tend simply to give way to the more dominant birds or family groups within the flock, this final 'nuclear option' is generally only reached when opposing parties are well matched. Submissive behaviour will usually resolve the matter and involves the defeated bird turning away with its bill tilted up so it can clearly guard its rear while beating a hasty retreat. Any successful aggressive encounters are usually followed up with 'triumph' displays, in which the winning swan will return to his mate and cygnets, before displaying with a neck stretching, wing flapping and loud honking routine as his family enthusiastically joins in the celebrations. By always keeping no more than just a couple of wingbeats apart when in amongst the wintering flock, the dominant families will always have muscle on hand to take the pick of the best foraging and roosting spots for the entire duration of the winter. In fact at locations like Slimbridge, where many of the swans will return each winter, this hierarchy can become so entrenched that dominance will frequently be held from year to year.

Freshly arrived from northern Europe, any aggression between individual Waxwings from the same flock only ever seems to be minimal as these gentle birds adopt a far more consensual approach to feeding than the obstreperous Bewick's Swans. Where Waxwings go after their initial arrival is now more clearly understood thanks to work carried out by colour-ringing studies. By ringers using different colour combinations on the legs of each Waxwing netted, birdwatchers up and down the country are then able to report the location of different individuals without the need for their recapture. During the influx of 2010/11, for example, the Grampian and Orkney Ringing Groups managed to catch and individually colour-ring almost 500 Waxwings through late October and November in an attempt to try and track the movement of the birds as they spread across Britain. With competition for berries patently at a premium, Waxwings quickly departed both ringing locations, with one particular Orkney-ringed bird rapidly relocating to Norfolk and a number of

Aberdonian-ringed Waxwings subsequently reported from Dunfermline, Lothian, Glasgow, Cumbria and Manchester. The subtleties of each Waxwing invasion will of course play out differently, but the general pattern of movement usually sees birds streaming in a southerly and westerly direction as the food becomes stripped out of their original entry points. Interestingly, one of the Waxwings caught in Aberdeen was an adult male that had been ringed in Svartbyn, a town in northern Sweden, close to the Finnish border. After having travelled 1,654km all the way to Duthie Park in Aberdeen, this bird then sadly proved to be a window casualty just a week later, but only after having served as a standard bearer to further illustrate the strong Waxwing link between Sweden and Scotland.

The Swallow's link between Britain and South Africa was first established over a century ago, when a female ringed in Staffordshire in 1911 was subsequently recovered from the Province of Natal in December of the following year. This was merely the first of an astonishing 447 British-ringed birds that have so far been recovered in South Africa, patently making the southern reaches of the African continent their destination of choice. It's believed that most returning Swallows will generally be faithful to their winter quarters, although locations can vary if the weather impacts on food availability. Just as in Britain, the Swallows will forage across a wide variety of habitats, as they swoop over grassland, forest edge, cultivated fields and particularly any wet environment, before then returning each evening to their communal roosts.

Up to 2011, not a single ringed British-breeding Nightingale had ever been reported south of the Sahara, leaving the sum total of knowledge about the location of 'Nightingales in November' to be collated from the information gleaned by the geolocator attached to 'Nightingale OAD'. As geolocators have an inherent degree of inaccuracy in the tropics, where dawn and dusk vary little throughout the year, the BTO researchers interpreting the data believed OAD most likely to have been passing through Mauritania in early November. Comparable in size to Egypt, around 90% of this huge but impoverished country is categorised as 'Sahara Desert', with most of the population living in the south-west, which experiences a marginally higher rainfall than in the desert interior. Despite a substantial mineral wealth consisting of significant deposits of iron ore and oil, severe international sanctions following a coup d'état in 2008, a poor human rights record and corruption have all combined to ensure that Mauritania remains a desperately poor desert nation.

The topography of the country is generally considered fairly flat, with vast arid plains broken by occasional ridges and

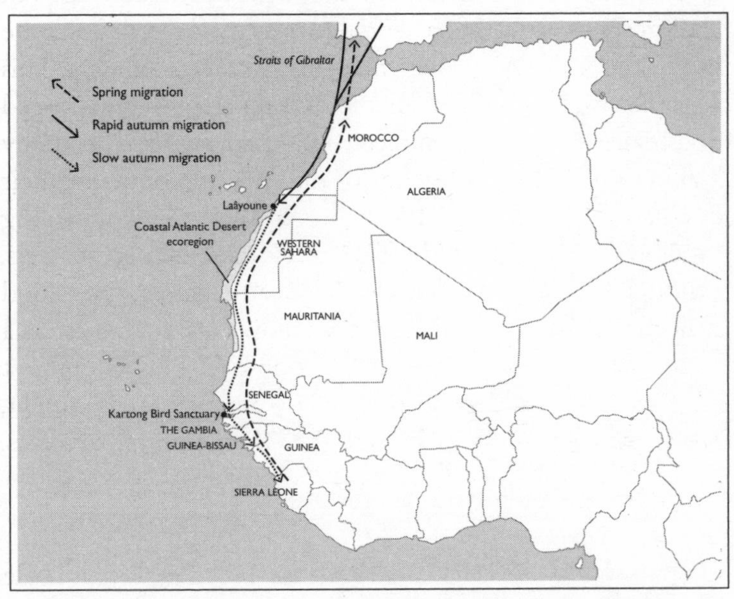

The believed autumn and spring migratory routes of Nightingales between their stopover sites in southern Europe and overwintering destinations in west Africa.

cliff-like outcrops. To the west, between the Atlantic Ocean and the higher desert plateaus, the land alternates between clay plains and sand dunes, or ergs, some of which can become mobile during high winds. Following the Atlantic Coastal Flyway, which is considered a key route for many migratory birds moving between Europe and Africa, the Nightingales in early November are still believed to be steadily working their way along this route, represented along this particular stretch by the 'thin green line' of the Atlantic coastal desert. As the birds advance further south, their move into the semi-arid savanna close to the border with Senegal will bring a measure of relief for the Nightingales, as the marginally higher annual rainfall closer to the equator enables the desert scrub to become steadily replaced by wooded grassland and bushland.

Over 4,000km to the south-east, the annual rainfall of around 180cm experienced in the Congolian swamp forests will find the Cuckoos experiencing a somewhat different climate to the Nightingales currently working their way along Mauritania's arid coast. Undoubtedly keeping a low profile in these wet, humid forests, the Cuckoos will not be the sole British-breeding representative to spend winter in the heart of Africa, as Swifts have recently been revealed through geolocator technology to spend most of their winter in the Congo too. Adopting a somewhat higher profile - and niche - than the Cuckoos, the Swifts will spend most of the winter months sailing above the canopy as they take advantage of the abundant insect life emanating from these incredibly diverse forests.

Certainly the contrast in habitat type could not be any greater than between where Cuckoos and Puffins will choose to spend their winter. Weighing in at just under 400g, and only marginally larger than a pint pot, most British-breeding Puffins at this time of year will be at the mercy of the North Atlantic as they spend most of the daylight hours diving for dinner. How far Puffins disperse after leaving their breeding grounds was very poorly understood until Mike Harris's team on the Isle of May began to follow the movements of 13 birds with the help of geolocators. The first revelation was that ten of the tagged Puffins, upon leaving their breeding colony on Scotland's east coast, didn't stay in the North Sea, but immediately travelled anti-clockwise around Scotland's north coast to head out into the Atlantic Ocean. At some point during the winter all Puffins will spend four to five weeks with their wings effectively clipped as they moult their primaries, but this didn't restrict one tagged individual, 'bird 6539', travelling an estimated 3,000km during an extensive tour of the north-east Atlantic during the winter months. Initially heading up to the waters around the Faroe Islands, this bird subsequently travelled down to the west of Ireland in early November, before then returning further north, presumably after having moulted its primaries, to spend time off St Kilda. Although many of the Puffins followed did not travel as widely as this adventurous individual, it's likely that many birds may well be covering a much wider area of ocean during the winter months than was previously thought.

As Puffins undertake expansive voyages out at sea, the adult Tawny Owl's entire universe will be the territory currently being vociferously secured for the winter and the breeding season beyond. The size of territory tends to depend on the quality of the habitat, with an established pair in mature, high-quality woodland often possessing a much smaller piece of real estate compared to those Tawnies living in either open woodland, or farmland containing small isolated patches of forest. In studies made in and around the intensively researched Wytham Woods near Oxford, the Tawny Owls living in the deciduous woodland were found to occupy a territory, on average, of around 18 ha, while pairs living nearby in mixed farmland needed a territory of around double the size. Clearly, in the world of the Tawny Owl, territories are all about quality not quantity. Additionally, the boundaries between prime territories can often stay remarkably stable between years, even remaining fixed despite changes in ownerships as established birds either die of natural causes or become ousted by any young pretenders waiting in the wings.

With Robins also considered to be strongly territorial at this time, it is perhaps a surprise that Robins have been found to roost communally at some locations. After a roost was discovered in Aberdeen, further searching revealed five more around the city, with the largest gathering holding 53 different individuals throughout the season. By netting and colour-ringing birds using these roosts, the Robins were tracked down to breeding territories in local woodlands, which were patently exchanged in the autumn for different territories in nearby housing estates. While yielding more than enough food throughout the winter, thanks primarily to the generosity of householders leaving out food, it seems these urban territories may not be in a position to provide a sufficient number of roosting sites, hence the reason for the birds grouping together. Communal roosts have also since been located in Cambridge, where it was thought that many of the participating Robins were young birds unsuccessful in securing a territory in the autumn. Using these communal roosts may enable any newcomers to get the feel of the local situation on the ground, empowering them either to stay on in the hope of ultimately gaining a territory, or simply moving on elsewhere if they considered the surrounding areas overly congested.

Keeping rigidly on territory wherever possible, each Kingfisher will defend its patch throughout the winter as if its life depended on it, which for once is not hyperbole. Certainly in the breeding season, territories can vary from a couple of hundred metres to a few kilometres of riverbank or lake perimeter in length, but winter territories generally tend to be smaller. Within that territory, most of the birds' activity will in fact be mostly confined to an even smaller core area, containing both good fishing spots and suitable roosting locations, as they knuckle down to the serious matter of seeing out the winter.

More confined to a home range rather than a territory, which they'll share with a motley band of other bird species by day, Blue Tits will normally prefer to spend their nights in a far more solitary fashion. Having such a small body mass can make surviving the long winter nights a tough ordeal, so the Blue Tits' technique for keeping themselves warm at night involves finding accommodation in the form of sheltered nooks and crannies. Keen to conserve energy by reducing heat loss means these tight spots may well be at a premium, with the result that any Blue Tit already occupying a favoured location as dusk approaches should easily be able to drive away any other bird attempting to gatecrash this party for one.

Due to the abundant food on tap in towns and cities, many urban Peregrines will attempt to stay on their territories all year round, but certainly those Peregrines at traditional rural sites in remote Scottish or Cumbrian fells, for example, may well be forced to relocate for the duration of the winter as their prey seeks more clement conditions elsewhere. Often these tough northern birds may opt to station themselves at, or close to estuaries, where bird numbers will have been steadily building up throughout the autumn. Originating from breeding areas as diverse as Siberia, northern Europe, the Russian Federation, Iceland, Greenland and north-east Canada, literally millions of wildfowl and waders are attracted to British estuaries each winter, thanks to a combination of our mild maritime climate and large tidal ranges. Feeding on the extensive areas of mud and saltmarsh around our coasts as the tides ebb and flow, large flocks of Teal, Dunlin and Knot will need to keep their eyes peeled if they are to avoid appearing on the diet of this most ruthless of avian predators.

Mid-November

Having been on the move ever since exchanging Europe for Africa a couple of months earlier, British-bred Nightingales should finally be arriving at the destination that will become their home until at least Christmas. Even before the wintering location of Nightingale OAD was revealed by analysis of the data from its geolocator, it had long been suspected that British Nightingales spend at least part of the winter in the countries of Senegal and The Gambia. Both situated in West Africa, the much larger country of Senegal is externally bounded by the Atlantic Ocean to the west, Mauritania to the north, Mali to the east and Guinea and Guinea-Bissau to the south. Senegal also almost entirely surrounds the Gambia – mainland Africa's smallest country, which apart from a short stretch of coastline bordering the

Atlantic Ocean, is a country consisting of little more than a long strip of land straddling the Gambia River.

Both Senegal and The Gambia are considered to be politically fairly stable, with the former's economy driven by exploiting and refining natural resources, while the latter's is dominated by farming, fishing and tourism. The region is characterised by a tropical climate, with a mostly pleasant to stifling heat throughout the year and well-defined wet and dry seasons. The vegetation close to the coast, in both countries, mostly consists of savanna woodland, thorny scrub, gallery forest and wetland, with The Gambia being particularly well known amongst the British birdwatching community, enticed by an impressively high avian checklist for such a small country. Highlights of a trip to The Gambia in the winter will include many species never likely to be seen in north-west Europe, together with a smattering of visitors only familiar to Britain during the summer months, such as Ospreys, Sandwich Terns, Reed Warblers and of course Nightingales. Just as in their English breeding grounds, when in Senegambia the Nightingales tend to stay true to their naturally skulking behaviour, preferring to keep low down in the tangled, thorny savanna scrub typifying their accommodation at this time of year.

Having completed most of their moult before they left British shores, any Nightingale glimpsed should be looking decidedly fresh as it settles down into a daily routine. The timing of the Nightingales' moult is thought to contrast starkly with that of the Cuckoos, which are only believed to carry out most of their moult when well ensconced on their Congolian wintering grounds. Due to the Cuckoos' effective disappearance from view for over 80% of the year, their moult is little understood, but it is believed that a proportion of adults may well replace at least some of their body, flight and tail feathers before crossing into Africa. With the on-board satellite tags revealing that the Cuckoos move little at this time of year, it is reasonable to suggest they will be using this settled period to actively moult their remaining flight feathers. Cuckoos possess ten primaries, with ringers identifying the innermost primary (closest to the body) as P1, while the outermost feather is P10. These ten primaries are believed to moult in two series, with feathers P1 to P4 shed in descendant fashion – or away from the body, while P5 to P10 are moulted in an ascendant manner – or towards the body. To further complicate the issue, P5 to P10 moult in alternate fashion, giving a possible replacement sequence of 9-7-5-10-8-6. This regular alternation between growing and non-growing feathers should still presumably give the Cuckoos enough lift to both evade predators and forage effectively for food.

Amongst Britain's summer visitors there seems a distinct pattern, with those species migrating the furthest south generally delaying the majority of their moult until reaching their wintering grounds. Like the Cuckoo, this delayed moult will also occur with the Swallows currently entrenching their positions in and around South Africa's Western Cape. Despite the Swallows' moult having started around the beginning of October, or even earlier in a few cases, the entire process of replacing around 1,500 feathers may well take as long as five or six months. This protracted process could mean that the large outer primaries and tail feathers in many Swallows might still be actively growing upon their arrival back in Britain. For the Bewick's Swans currently bedding in to their wintering grounds, the moult of the adults' main flight feathers will already have been completed before they migrated south, leaving the exchange of the best part of 25,000 body feathers to occur at a more leisurely rate right the way through to February.

During their marathon flight to Britain the Bewick's Swans will have lost considerable condition, with newly-arrived males weighing in at little more than 5kg and the females around a kilo lighter. In addition to sorting out their hierarchy within the flock, piling the lost weight back on will also be undertaken as a matter of some urgency. Certainly around the WWT Slimbridge reserve in Gloucestershire the swans are thought to initially prefer feeding on the rich improved pastures around the Severn Estuary. However, for those birds returning to the Ouse Washes in Cambridgeshire, or WWT Martin Mere up in Lancashire, the habitat of choice for putting the pounds back on will be a whole variety of arable fields. In those areas where feeding amongst arable crops is favoured, the numbers of Bewick's Swans tend to be initially highest on the stubble fields where the birds are presumably mopping up any spilt grain. But as soon as this resource starts to run out they will then move on to polishing off any waste from the remains of root crops left in the fields after harvesting. So nutritious are these varied foodstuffs, that scarcely a month after having returned back the males should have been able to quickly increase their weight back to a fighting fit 7kg, while the females will end up a touch lighter, at around a kilo less.

Tipping the scales at a mere 55g, or just under 1% of the weight of an average Bewick's Swan, the Waxwing's appetite could be considered even more prodigious in proportion to its size, which will primarily be due to the relatively poor nutritional value of many berries. Even in invasion years, such as the last one experienced in 2010, there still seem to be surprisingly few birds reported south or west of a line drawn between the Isle of Man and London by the middle of November. This means that birdwatchers in cities such as Brighton and Bristol may just have to show a touch more patience yet before the first of these delightful invaders finally begin turning up on their doorsteps. Following the movement of close to 500 Waxwings that were individually colour-ringed by the Grampian and Orkney Ringing Groups in 2010, it appeared that many of the birds quickly relocated in southern England were individuals that had patently leap-frogged those colour-ringed Waxwings which had already moved down into northern England from their original ringing locations. Presumably, as the best food sources become steadily cleared out in a rolling wave across the country, 'getting ahead' is the smart way to ensure access to fresh, untapped supplies.

While the Waxwings' winter diet is laid out for all to see, trying to elucidate exactly what food Puffins eat when dispersed out to sea is decidedly more difficult. What little information there is on their North Sea diet comes from 68 dead birds found washed up on beaches. In two-thirds of these, remains of gadoid fish (the fish family containing Cod, Herring and Pollack), Sprat, Goby, Three-spined and Fifteen-spined Sticklebacks and Snake Pipefish were all found. In just under half, the jaws of polychaete marine worms were found, and from one dead Puffin a squid beak was recovered. These remains could have been a slightly misleading representation of their true diet, however, as bones of fish, plates of pipefish, spines of sticklebacks and jaws of marine worms will undoubtedly persist in the stomachs of the birds longer than small, soft-bodied prey items. More accurate information on the Puffin's winter diet may have come from birds which were unfortunately snared by fishing long-lines between

Norway and Iceland. On analysing the stomach contents of 11 accidentally hooked birds, researchers found that they had been feeding on Glacier Lanternfish, squid and polychaete worms. While it is striking that marine invertebrates undoubtedly form at least part of many Puffins' winter diet, in addition to a whole variety of fish species, educated guesswork would suggest that when the birds are foraging below the surface, there may well be more of an opportunistic element to what they are catching rather than the precise targeting of any one species.

Thanks to recent work carried out by Peregrine experts Ed Drewitt and Nick Dixon, the diet of urban Peregrines throughout the year is now known to be far more diverse than just Feral Pigeons. By collecting and identifying the prey remains from Peregrines in the cities of Bristol, Bath and Exeter between 1998 and 2007, they were able to identify an astonishing total of 98 species. Varying in size from the diminutive Goldcrest to the substantial Mallard, this emphasises not only the huge diversity of prey now taken, but also the opportunistic nature of the modern urban Peregrine. In their study Feral Pigeons were still identified as the most important prey species, comprising 42% of all items taken throughout the year, but certainly in November, a massive spike was recorded in the numbers of Common Snipe, Woodcock, Fieldfare and particularly Redwing that were taken. All primarily continental breeding species that visit Britain in large numbers during the winter, these birds are almost certainly being singled out while on passage at night.

Despite Robins eating a diet largely based on insects and arthropods between spring and autumn, vegetable food in the form of fruits, seeds and nuts will inevitably become an increasingly important component as the temperature drops. During early winter, fruits taken from both gardens and the wild will include currants, apples, blackberries, haws, rose hips, elderberries and Yew berries. Having learnt to also take advantage of any food we put out will undoubtedly boost their chance of survival, but in those years with a harsh winter their mortality may still be extremely high. From studies it seems that it is the length of the cold spell rather than the intensity that seems to cause the Robins most problems. Weighing in at around 15 or 16g in the summer, a fat layer put down before the weather turns cold will see the Robins a decidedly chunkier 22 or 23g during the height of winter, which can then be quickly metabolised to help tide them over any short periods with little or no food. However, an analysis of ringing recoveries during harsh winters has shown that most Robins are usually recovered dead in the second week of a cold spell, doubtless birds with their fuel tanks already having been drained dry after the first seven days.

Either unwilling or unable to move out of their immediate territory, as the temperature begins to plummet, Tawny Owls will need to rely on both an intimate knowledge of their local patch and resourcefulness to see them through the winter. Those Tawny Owls holding territories in more rural locations will undoubtedly rely mostly on Wood Mice and Bank Voles throughout the year, while urban pairs may take a larger proportion of birds. However, if push comes to shove, these adaptable and opportunistic birds will also consider taking more unusual prey. Grey Squirrels, for example, have occasionally been recorded at a variety of locations, Edible Dormice have also been documented being taken by pairs holding territories in the Chilterns and even the remains of fish have been found in some Tawnies' diets. With a far wider prey spectrum than all the other British owls, it seems that, in essence, if it's edible then the Tawny will eat it!

Unlike our mostly sedentary Blue Tits, which will frequently stay within their home ranges for the duration of the winter, migration amongst continental Blue Tits is a far more commonplace activity, as birds nesting at higher latitudes are forced further south to escape the severe north European winter. In fact the only significant movement of any British-breeding Blue Tits will be from those birds nesting in more northerly or upland sites, which will have little choice other than to accept retreating to more benign locations or adjacent low-lying areas as conditions deteriorate. Being such a popular species for research, with close to four million individuals having been ringed, the huge dataset managed by the BTO indicates that females tend to disperse further afield than males. One possible reason for this differential movement between the sexes may be down to an element of competition, with the larger, more dominant (and frankly unchivalrous) males effectively ousting the females from the best neighborhoods.

Likewise with Kingfishers, any birds which bred at higher altitudes may well have already moved towards the lowlands and coast where new winter territories will then need to be established. Kingfishers have not traditionally been as extensively ringed as many other species, but the scant data available suggest that most northern Kingfishers will still be reluctant to move any more than 15km from their breeding territories. However, with nine Kingfishers initially ringed in Britain and then subsequently found abroad, there does seem some evidence of a migration south and west, with a few individuals even recorded to have taken advantage of the milder winters experienced in countries such as France or Spain.

Unable to cope with any prolonged cold snap, the option of making a quick bolt for mainland Europe's Atlantic coast as soon as the weather turns unpleasant appears a much more commonly exercised option for the Lapwing than for the Kingfisher. Out of 845 Lapwings ringed in Britain that were subsequently recovered abroad, an astonishing 714 were reported from France, Spain and Portugal. This high total suggests that the warming effect of the Gulf Stream along Europe's western seaboards should offer frost-free feeding, which will more than compensate for the effort and inherent risk of undertaking the journey in the first place.

Late November

With autumn drawing to a close it seems the timing of the Kingfisher's breeding season back in the summer will play a large part in dictating when the adults manage to complete their annual moult. Those birds, for example, that concluded their breeding cycles by mid-August should by now be looking particularly smart as they see their moult reaching completion. However, any individuals still in mid-moult due to a late-fledging brood may be left with little option other than to suspend this energetically demanding process to a time when their food supply improves. With the larger, slower-growing primaries usually being retained, these feathers will then be the first to be replaced upon recommencement of the moult cycle after the following year's breeding season. This suspension will have the knock-on effect of handicapping some Kingfishers due to the moult of one wing being out of synch with the other and even an unbalanced tail too, all factors which could limit their ability to see out the winter. Any surviving juveniles still going strong by this stage will also have completed a partial moult of their entire body and certainly most, if not all, of their tail feathers. Ringers report some youngsters' moults can even drag on into December in those fledging from particularly late broods, but the all-important flight feathers, which initially emerged just a couple of weeks after hatching, will now not be replaced until after their first breeding season.

Having started their annual moult back in their taiga forest breeding grounds as long ago as early August, those adult Waxwings reaching Britain should also be looking bright and fresh as the last of their new set of feathers reaches full length. Despite undergoing a partial moult, those surviving youngsters from the summer's broods will not develop their full adult plumage until well into the next year. The retention of their original flight feathers for at least the next eight months will mean these first-year birds will not have been able to develop either the pronounced yellow on the outer webs of their primaries, or as many prominent red waxy tips on their secondaries. These waxy tips are not only the feature that gives Waxwings their name but are also the badge of status for any birds reaching adulthood. Despite the oldest known Waxwing having survived to the grand old age of 13, it is highly likely that the average bird's life expectancy will be distinctly less. With over 3,100 birds ringed across Britain, only seven Waxwings have been recorded returning here in subsequent winters, and with mass invasions being irregular at best, this would suggest that most individuals may not reach Britain any more than just once in their lifetime. This will make their trip all the more remarkable as they take in a part of northern Europe that many of the flock will have never seen before.

Like the juvenile Waxwings, retention of the young Lapwings' flight feathers until after their first breeding season will mean that when observed in late autumn an experienced eye should still be able to pick out the immature birds among the mixed-aged flocks currently dotted around the British countryside. As Britain has been subjected to a recent run of comparatively mild winters, the Lapwing seem to have responded with a shift towards wintering in the traditionally colder, but possibly more productive feeding grounds in eastern England. Feeding on the rich arable fields, which form such a dominant part of the agricultural sector in this part of the world, the Lapwings' mobile nature should enable them to quickly respond if freezing temperatures suddenly make feeding conditions decidedly more difficult. Moving south and west, the Lapwing will then often target those habitats less prone to frosts, such as pasture and grazing marsh.

For young and inexperienced Blue Tits in particular, summer and autumn can be a period of peak mortality as they run the dual risk of predation or starvation. However, for those youngsters surviving this brutal examination, and along the way acquiring both a thorough knowledge of their home range and far better insulated body plumage, late autumn should finally see their chances of reaching next spring begin to improve considerably.

Likewise for any juvenile Bewick's Swans which successfully fledged from their predator-riddled and climatically unpredictable breeding sites up on the Arctic tundra, the British wintering grounds will represent a place of comparative sanctuary. The young Bewick's Swans' survival prospects will also be enhanced considerably by the constant company of their supremely experienced parents, many of which will know the lie of the land at their winter quarters intimately.

Extensive research carried out at Slimbridge Wetland Centre in Gloucestershire has revealed that Bewick's Swans are surprisingly long-lived birds, which not only need a number of years to pair and breed, but also have a low reproductive output each year. Most of the swans only breed successfully for the first time when between four and six years of age, but this can often occur far later, such as the male swans 'Victory' and 'Money', which were not recorded with cygnets until ten and eleven years of age, respectively. The Bewick's Swan is of course strongly monogamous by nature, with experience often being the vital ingredient dictating reproductive success, but even amongst long-standing pairs, those seasons when young are produced can frequently be little more than erratic at the very best. Of 27 pairs followed at Slimbridge which were known to have been an item for at least ten consecutive years, only seven were found to have consistently and successfully bred, with young present, on average, at least one year in two. In fact, the best breeding rates were achieved by just two pairs – 'Dougie' with 'Estralita' and 'France' with 'Valois' – which both successfully managed to breed in nine out of 13 winters during the study period. Not only will breeding success vary widely between pairs, but also of course between years, with poor breeding seasons seeing only around 4% of the flock represented by juveniles, as opposed to about 20% of the flock being cygnets when conditions up in the Arctic have been far more favourable and predators less effective.

By contrast to the sociable nature of Bewick's Swan families, any Robin holding a winter territory will have no sentimental thoughts whatsoever towards any of its other family members as it works hard to maintain an exclusive territory for one. Certainly in David Lack's pioneering study on Robins in Dartington, Devon, the majority of individuals which established winter territories in his survey area were male, with only a quarter of females reckoned to have stayed throughout the entire year. He believed the majority of his female Robins, along with a smaller proportion of first-year and older males, disappeared for a large part of the winter. They would then arrive back early the following year, to either pair up with a resident male or, in the case of the migratory males, to try and carve out a territory for themselves. Lack was subsequently proved correct in his assumption, as ringing has since revealed not only a substantial number of British-bred Robins opting to relocate elsewhere in Britain, but also a smaller figure preferring to spend their winter on the continent. The three countries accounting for most British-ringed Robins are France, Spain and the Netherlands, all countries influenced by the warming effects of the Gulf Stream along the Atlantic seaboard. Ringing in Spain has additionally shown that some Robins of unknown origin will return to the same Spanish territory in successive winters, raising the intriguing possibility that some Robins may spend their lives switching between a breeding territory in Britain and a winter pied-à-terre on the continent.

Robins are not the only bird from our chosen twelve to sing for their supper during the winter months, as the British-breeding Nightingales, currently overwintering in west Africa, can also occasionally be heard uttering snatches of song from deep within the Senegambian scrub at this time of year. Unlike in England, where, upon arrival, they will be initially keen to sing both during the day and night, in Africa their performances are usually reserved for just the matinee slots, with the song also tending to be shorter, less complicated and more fragmented. Exactly why the males sing on their wintering grounds is not entirely clear, but there may be a territorial element, or it could just represent the perfect opportunity to practise their bewilderingly complicated song, in preparation for their return back to the breeding grounds. By contrast, all British Cuckoos, currently centred around the Congo River and the surrounding Congolian swamp forests, are thought to make little, if any, noise in their wintering quarters. Rarely even observed in this vast and largely undisturbed habitat, it would seem that quietly moulting and feeding out of sight of the innumerable Congolese predators may well be the best strategy for these highly secretive and mercurial birds.

Also keen on keeping a low profile, any adult Peregrines not actually out hunting at this time of year will be keen to spend as much time as possible conserving energy, which will simply be achieved by discreetly perching out in a sheltered spot. Providing they are able to find enough food to keep their internal furnace fuelled, their freshly moulted feathers should at last be in a position to keep them sufficiently warm irrespective of what the British winter dishes out.

Unlike urban Peregrines, which are able to hunt around the clock in winter, the Puffin's inability to hunt at night will mean its days during winter will be kept incredibly busy if it is to find enough food to survive in this hostile environment. Using their short, stubby wings to propel them below water, it was widely assumed from the limited data gathered that Puffins routinely dived to substantial depths in pursuit of food. A depth of 68m, for example, was recorded off Newfoundland, with Puffin expert Mike Harris initially estimating average depths to range from 21m to 33m. However, by attaching time-depth recorders to Puffin leg-rings, Harris's team were subsequently able to gain a much more accurate picture as to both the depth reached and the time the Puffins spent under water. The research revealed that although the Puffins do commonly go down to 35m when pursuing prey, the depth when all dives were combined averaged at only around 4m. What fascinated the researchers, however, was not only how shallow their study birds went, but how often they dived, with the key technique seemingly being 'little and often'. One bird followed in their study, for example, made 194 dives in just 84 minutes, with an average dive lasting just 28 seconds, which gave just six seconds of recovery time at the surface before submerging once again. Another bird followed made 442 dives in just over two and a half hours, in the process spending 70% of this time underwater. Doubtless, as the days continue to shorten towards the winter solstice, just finding enough to eat could well occupy virtually all their waking hours.

Also confining their hours of feeding to between dawn and dusk, the British Swallows currently overwintering in the

Western Cape of South Africa will converge at favoured winter roosts from as far away as 50km as the light begins to fade each day. As Swallows prepare to depart our shores in the autumn, reedbeds are often considered the prime roosting habitat of choice. However, in Africa, in addition to wetland sites, the Swallows will also regularly roost amongst sugar cane, maize and even trees. Some of the best-known sites will be occupied each winter, with certain roosts known to have been used for at least 50 years and possibly containing over a million birds, while others are much smaller and ephemeral by nature. As Swallows begin to arrive at the roost from all points on the compass, initially their technique seems to involve little more than flying around aimlessly, but as the light level drops, presumably below a certain threshold, they will then form tighter flocks as they rapidly wheel over the roost. As the Swallows continue to swirl directly above their accommodation, small groups will then begin to peel away from the main flock before dropping vertically down into the vegetation below. Presumably this method of synchronised flying, as with Puffins approaching dry land early in the breeding season, will make it more difficult for raptors to target and catch individual birds. Taking a while to settle down for the night, there may also be some shifting of position, together with accompanied twittering, as juveniles are shoved out of the best spots by older birds. In spite of realising the benefits of a safety in numbers strategy at night, Swallows are rarely tolerant of close contact with each other and will try to keep a respectable distance wherever possible. Possibly the only time when their personal space is invaded will be on those nights when the weather has turned so cold that communal warmth will trump their naturally frosty attitude.

In the particularly large wintering roosts the majority of birds tend to be juveniles and this proportion will become even higher as the austral summer proceeds, possibly even approaching close to 100% by the time the older birds have started to return to their breeding grounds in the New Year.

As the Swallows seek each other's company at night, late autumn should also see any established pairs of British Tawny Owls finally roosting closer together again for the first time since the summer. The number of roosting sites will also be narrowed down to a few favoured locations as the temperature continues to fall, and safe in the knowledge that their territory should be fully secured, the pair will finally be able to turn their heads slowly towards the breeding season early next year.

December

Though it's tempting to 'hibernate' indoors during the month of December, a festive walk on the wild side could be rewarded with surprising views of animals, as the short, cold days see them dropping their guard to forage out in the open. Midwinter is certainly not downtime for the Red Fox, for example, as the males will be vocally and actively defending their territories with the mating season just around the corner, and garden bird-feeders across the land will be bustling with business too. Freshly moulted, the tits and finches will be a particularly active presence in virtually any garden, and with most of our summer visitors happily settled in Africa, our resident Robins and Tawny Owls will have scarcely budged all year. Throughout this whole month, the birdlife eking out a living in frosty Britain will have only three things on its mind: keeping well fed, keeping warm and keeping safe.

Early December

Having already moulted their flight feathers by late summer, it could take until early December before adult Tawny Owls will feel sufficiently well equipped to repel the worst excesses of winter and finally replace the last of their body plumage. Tawny Owls are one of very few British species to exhibit more than one colour form or morph, with the majority of birds' plumage split into one of two categories – either principally brown or mostly grey. Certainly in Britain, the brown colour morph is by far the most common, which contrasts with Finland, where historically grey Tawny Owls have tended to predominate. Plumage colour is believed to be hereditary, with a long-term study of Finnish Tawny Owls finding that not only was the grey colour genetically dominant over its brown counterpart, but in particularly severe winters with thick snow cover, the grey-coloured birds had a higher survival rate. However, the milder winters over the last 28 years in Finland have seen the proportion of brown Tawny Owls increasing from 30% to 50%, a possible case of evolution driven by rapid climate change. With the brown owls seemingly less well adapted to colder conditions, this may indicate why in Britain, with its relatively benign winters, the majority of Tawny Owls will indeed continue to stay tawny.

Quite literally half a world away from the Tawny Owls, it will be months before the British Swallows currently overwintering along South Africa's Western Cape will see their plumage once again at its brilliant best. Becoming progressively warmer with each passing day as the austral spring takes hold, the average temperatures of around 20°C around Cape Town in early December should see plenty of flying invertebrates for the Swallows to feed on as they trawl the skies from dawn to dusk. Studies have indicated that the overwintering Swallows tend to forage mostly within 50km of their favoured roost, possibly only changing where they'll spend the night when food becomes scarce.

Likewise, those Cuckoos encountering good feeding opportunities in the Congo may well be reluctant to undertake any large-scale movements for the rest of the winter. An exception to this largely sedentary rule, however, was provided by Chris the Cuckoo, the only bird to have been successfully tracked for four successive winters by the BTO using satellite technology.

First tagged with a transmitter in Norfolk in June 2011, Chris was subsequently followed for four 'African tours' until his sad demise while crossing the Sahara in the summer of 2015. With most Cuckoos safely ensconced in the Congolian swamp forests, during early December in both 2013 and 2014 Chris suddenly upped sticks, before flying 800km further south to Angola. Around twice the size of France or Texas, Angola is bordered to the south by Namibia, with Zambia to the east, the Democratic Republic of Congo to the north-east and the southern Atlantic Ocean to the west. The country has vast mineral and petroleum reserves, but the economy is only just recovering from an intense civil war that lasted until 2002, and which has ensured that Angola still retains one of the worst life expectancy and infant mortality rates in the world. Like the rest of tropical Africa, Angola experiences wet and dry seasons, with a short rainy season between February and April, a dry period between May and October and intermittent rainfall during the winter.

As a proportion of the Cuckoos proceed down to Angola, most British Nightingales are currently assumed to be still holding patterns, and small territories, in the coastal scrub dotted along West Africa's coast. Experiencing a dry season from November to mid-May, yet with temperatures still likely to be above 20°C, it is highly likely that western Senegal and The Gambia will be coming progressively drier as the month proceeds. The weather in the North Sea and north-east Atlantic Ocean, however, will be a touch more unpredictable for the overwintering Puffins. With these hardy little seabirds currently scattered across a whole range of sea areas, such as Forties and Dogger in the North Sea, to the Faeroes and Southeast Iceland north-west of Britain, and Rockall, Bailey and Shannon to the west, the Puffins will have few opportunities to shelter from any gales or storms which appear with monotonous regularity at this time of year.

Severe weather on the continent will also have a direct impact on the numbers of Bewick's Swans overwintering in Britain. Any freezing conditions in the Netherlands, for example, where up to 70% of the total population of all Bewick's Swans in north-west Europe can be held in early winter, will suddenly result in Britain seeing an influx of 'new' birds moving away from ice-bound lakes and frozen fields. Capable of making the journey in one short-haul flight, it won't be long from when the swans are forced out of their Dutch wintering sites to when they begin turning up in British overwintering flocks like WWT Slimbridge in Gloucestershire and WWT Welney in Cambridgeshire. Entering relatively settled flocks with an already established dominance hierarchy may initially result in aggression flaring up once again as the 'newcomers' fight for a grudging acceptance from the 'regulars'. Additionally, being probably unfamiliar with the best feeding and roosting sites means these Dutch birds may well have little choice other than to follow a watching and learning brief until they familiarise themselves with their new surroundings. Likewise, in a good Waxwing year, early December should see the number of Waxwings continuing to swell, as birds that arrived as early as October are joined by recent immigiants pushed across the North Sea as a direct result of a shortage of food in Scandinavia.

Systematically stripping the trees and berries as they merrily go about their nomadic existence, the Waxwings will usually choose to pluck most of the fruit directly from the shrub or tree, but will occasionally also descend to the ground at sites where food may have already ripened and fallen. Each berry is picked with a slight stooping motion and only held briefly in the bill before then being gobbled down following a quick flick back of the head. Being light and acrobatic birds means clinging to the undersides of branches to hoover up any harder-to-reach berries can be achieved with the minimum of fuss, and they will even briefly hover to pluck any fruit suspended from the extremities of the flimsiest twigs. While Hawthorn and Rowan berries are usually swallowed whole, the birds tend to be much fussier when feeding on Cotoneaster and Viburnum. These slightly less palatable berries will be dissected by their tweezer-like bills in order to extricate the seed and pulp, before the skin is then discarded. Any Waxwings feeding on large fruits, such as apples, will then proceed to apply a different technique again, using the bill like a dagger to liberate chunks of pulp.

As fruit is rich is sugar, but deficient in a number of other essential prerequisites, such as proteins, these frugivores will need to devour large quantities to garner sufficient nutrients for all their daily functions. The Waxwings cope with large quantities of fruit by having relatively large livers to help metabolise the excess sugar and will also need to drink frequently to counter the dehydrating effects caused by this incredibly specialised diet.

As Kingfishers' food doesn't grow on trees, in addition to maintaining an exclusive territory, they will also need a considerable degree of skill and level of luck if they are to catch enough food to see them through the dark days of winter. With only their own bill to feed, they can help their cause by spending long periods quietly perched out of view, thereby ensuring energy expenditure is kept to an absolute minimum. The Kingfisher's biggest enemy in the winter is prolonged freezing weather, as the formation of a thick layer of ice will instantly make fishing impossible. In very severe winters, many birds will simply abandon their territory for ice-free conditions near to the coast or head even further afield. A small number of Kingfishers ringed in Britain during the breeding season have even been recovered on the continent, with the highest number (three) being recorded from France, where the winter temperatures, certainly in the western half of the country, may well be a touch higher than across large parts of Britain. Whether these birds then return here when conditions finally began to improve is of course anyone's guess.

Being equally susceptible to freezing conditions and without a winter territory to tie them down, as soon as the temperature slips below zero many British-breeding Lapwings may simply either disappear off to the near continent, or at the very least head to the frost-free estuaries in the south and west of Britain. During intense cold snaps, movements of Lapwings can be quite dramatic, such as the 4,500 observed passing in a south-westerly direction over Tring, Hertfordshire in just 50 minutes on 9 December 1967. Once conditions do improve, however, the flocks are thought to favour a quick return 'whence they came', as competition for food in these hard weather refuges may quickly become intense as the birds are forced together in ever larger flocks.

As the days begin to shorten towards the end of the year, the more rural Blue Tits are thought to respond differently to the Lapwings by breaking up their large autumnal flocks into smaller and less mobile groups. In a long-term study of rural Blue Tits overwintering in an oak woodland in Kent, the individual birds seemed to spread themselves out across the wood, and confine their movements to within fairly small and defined areas. These wintering areas were not defended as strongly as individual breeding territories in the spring, but the boundaries seemed nevertheless to have been respected by neighbouring birds. By contrast, many suburban and urban Blue Tits seem to spend the winter roaming more widely than their country cousins as they take in a variety of feeding stations during the course of each day.

Harsh winters will inevitably exact a heavy toll on many of our smaller resident birds, but just like those Blue Tits preferring town to country, any Robin eking out a living alongside us humans will find its chances of survival hugely boosted by capitalising on any food we leave out. Historically happiest either taking food from a bird table, or cleaning up any scraps dropped by other more acrobatic birds helping themselves from feeders above, it seems that Robins have recently learnt an agility trick or two. A far more common sight in winter gardens these days is that of a Robin darting out of the vegetation to land directly on one of the feeder perches, before then grabbing a sunflower heart and quickly turning on its heels to consume its prize from nearby cover.

Recently voted Britain's favourite bird, the Robin undoubtedly achieved this spot in large part due to its endearing feature of being uniquely tame. Considered on the continent to be a shy woodland bird, in Britain it seems to revel in human company, leading to one of its commonest monikers, 'the gardeners' friend'. It seems the Robin perched on a fork handle is not just a concocted image to sell Christmas cards, but is down to the bird's genuine confiding nature, which it will use to take advantage of any morsels unearthed as we turn over our herbaceous borders and vegetable patches. It's believed this opportunistic action has been simply adapted from an ancient behaviour of following Wild Boar, a species which would have been common in prehistoric Britain. Guilty of disturbing the soil as they forage for roots, shoots, invertebrates and carrion, any food missed by the Wild Boar will represent a free meal for a sharp-eyed Robin following in close attendance. As Wild Boar became extirpated towards the end of the Middle Ages, the Robins may have simply looked upon us as little more than tall pigs!

Certainly for urban Peregrines, midwinter will also be a time for hunkering down in their respective territories as they go about the business of surviving winter. As the days continue to shorten, the birds will stay close to the core of their territory and even visit the nest site itself as they familiarise themselves with any scrapes used in previous seasons. This subtle shift from neutral to first gear will be the very first move towards the breeding season as they prepare to reassert their territorial rights and strengthen the pair bonds with their partner early in the New Year. On a day-to-day basis, however, most of their time will be spent roosting to conserve energy in between bouts either raiding their caches or hunting anything from pigeons to Pied Wagtails and Woodcocks to Water Rails.

Mid-December

When compared to deciduous woodlands, which are positively dripping with caterpillars in spring, most gardens are generally considered far poorer quality habitat for Blue Tits during the breeding season. However, as food supplies in wooded areas undoubtedly dwindle with Christmas just around the corner, garden feeding stations will suddenly become a much more attractive proposition to the local bird population. Frequently seen in ones or twos, it's easy to underestimate the number of Blue Tits visiting a popular feeding station, both during the day and throughout the course of the winter. But just how many different Blue Tits may drop in during a single day was not fully appreciated until intense ringing efforts were carried out in a series of gardens. The sheer number of birds netted was able to show that well-visited gardens are capable of pulling in as many as 200 different Blue Tits in a single day, as a whole series of small flocks move through on a 'regular beat'. In fact, garden birdwatching supremo Mike Toms from the BTO reckons that during the course of a winter, more than 1,000 different Blue Tits could visit a well-supplied bird table. This astonishing figure can easily be corroborated by BTO ringer Denise Wawman, who since 2008 has ringed 3,902 different individual Blue Tits in her small Minehead garden, with presumably many more individuals having avoided her mist nets altogether! Denise believes that many of the Blue Tits visiting her garden in winter hatched in local woodlands, with ringing recoveries showing that only a tiny percentage come from further than 5km away.

Unlike the mobile Blue Tits, those Robins holding territory will be keener than ever to stay close to home as they continue to eke out a living in wintry Britain. Any Robin zealously attempting to protect a patch will still be unable to prevent other individuals from trespassing all of the time, particularly in those 'hard to defend gardens' with plenty of food on tap. In fact, Denise Wawman has managed to ring a grand total of 362 Robins within the confines of her small, but well supplied Somerset garden since 2008. This fact alone suggests that the individual Robin that many people often mistakenly consider to be 'their own personal bird' may instead be a whole series of imposters either residing locally or just passing through. In particularly harsh winters, when survival becomes the order of the day, territoriality may simply break down at popular feeding sites, as the incumbent Robin realises that chasing away all other intruding birds is a waste of energy and a pointless exercise. Without any nesting Robins from year to year, Denise thinks of her garden as a 'neutral zone', where the Robins seeming to form a 'queuing system' for the bird table in very cold weather. She says 'they enter the garden from the north side, and fly across to the Lilac tree near the bird table, before moving through it towards the food. After briefly alighting on the bird table to grab a suet pellet, they then leave by flying around the house to the south. These departing Robins will then work their way back around the outside of the garden to rejoin the queue on the north side.'

Denise also adds that during the cold spells no fights were seen and there was only the occasional ruffling of feathers when a Robin went in the wrong direction or tried to queue jump!

Irrespective of the weather, those female Robins that moved elsewhere in Britain to maintain a different territory way back in the autumn, will from mid-December onwards be considering a move back to where they will intend to breed. Any females returning to a familiar area, upon choosing a mate will then quickly be pressed into service, as the pair work cooperatively to maintain a joint breeding territory right through to spring and beyond.

As the cold weather begins to bite, those territory-holding Peregrines which cached items earlier in the year when food was more abundant may well be forced into resorting to 'something they prepared earlier' to supplement their diet during times of hardship. With temperatures falling, the Peregrines can be assured that any Woodcock, Water Rail, Moorhen or Teal harvested during the autumn migration will suddenly see the shelf-life of their carcasses extended as the caching spots become turned into outdoor refrigerators. Sometimes these caches can run into tens of prey items and need to be stored with a degree of care to prevent them being discovered by other predators. Additionally, any items that become dislodged by wind or rain will not be retrieved if they tumble down to the ground below.

Unable to create their own food store for the lean times, any Kingfisher opting to stay within its own territory for the duration of the winter will have little choice other than to find enough food or suffer the consequences. Faced with freezing conditions, it's not unprecedented for Kingfishers to turn up at garden ponds hoping for an easy meal, and occasional observations have even recorded Kingfishers stealing food from Dippers and Water Shrews. In October 2007, one opportunistic or desperate Kingfisher in Essex was seen with a Pygmy Shrew in its bill, before the unfortunate mammal was presumably eaten. Such a highly unusual meal must be considered very much the exception to the rule for a bird that most consider to be almost entirely piscivorous, and usually when confronted with such untenable conditions on their own doorstep, many will simply move elsewhere.

After initially having turned up the volume in late October, presumably to prevent young Tawny Owls from attempting to settle within established territories, there will be a renewed urgency of calling by the adults as December proceeds. This 'second wind' will often continue right the way though into January and beyond as they gear up for the breeding season early in the New Year. For those pairs holding a rural territory, Wood Mice and Bank Voles are believed to form the most important component of their diet during winter, with a study between 1949 and 1952 in Wytham Woods, Oxfordshire revealing that these two species alone accounted for over 60% of all items caught during November and December. However, amongst those owls holding an urban dominion, birds appear to be the predominant prey items of choice all year, with a study in Holland Park, London revealing that birds formed 93% of the total live weight of all food caught by the owls. The main targets for these urban Tawnies seem to be House Sparrows, Feral Pigeons, thrushes, Starlings and Greenfinches. Despite the proportion of birds to mammals caught staying pretty constant throughout the year, the same study revealed that larger birds, such as Feral Pigeons and Jays, tended to be more commonly taken later in the year.

The type of invertebrates taken by overwintering Lapwings during the course of the seasons will of course be largely dictated by both the annual cycles of their prey organisms and the prevailing feeding conditions. The Lapwings' keen eyesight also means they need not just be restricted to feeding during daylight hours, as any nights brightened by moonlight will present excellent foraging opportunities, providing the ground isn't frozen of course. When not actively feeding or moving between sites, much of the rest of the time will be taken up by roosting. Preferring larger, more tightly packed flocks for roosting than when feeding, the sites Lapwings will pick for rest and recuperation tend to be primarily chosen with safety in mind. Roosting either during the day or night, many Lapwings will prefer either to use ploughed fields or tussocky grassland, which offer the dual benefits of shelter from the worst of the elements and concealment from predators. Overwintering flocks will also use flooded fields as roosts, which will be able to offer protection from any terrestrial predators not too keen on taking a dip, such as Foxes.

A recent phenomenon has seen Lapwings roosting – particularly during autumn and winter – on the roofs of industrial buildings in northern towns and cities. This was first noticed in Greater Manchester in 1984 and then subsequently recorded in and around Bolton, Bury, Oldham, Stockport and Wigan. In 1993 this behaviour also began to be reported from across the Pennines in cities such as Leeds and Bradford. Often comprising groups of up to 600 birds, most of these roosting flocks seem to gather on the roofs mostly after sunrise, presumably after a night's foraging, and then break up again just after sunset for feeding grounds at least 4km away. Remarkably, the roosting birds are surprisingly tolerant of all the activity that comes from residing in such close proximity to humans. Presumably once the birds have become accustomed to all the noise, then these industrial locations are in a position to offer some surprising advantages. From their lofty position these urban-roosting Lapwings will not just be out of reach of ground predators, but will also be afforded a level of protection from inclement weather – with the sloping roofs offering a variety of aspects for shelter and additionally the bonus of some free warmth transmitted through the roof from the building below. So, disturbance-free roosting right in the middle of our industrial heartlands might not be as daft as it sounds!

As the food becomes stripped out of the Waxwings' 'arrivals lounges' in the Scottish Highlands and Islands, which would have been thronging with Waxwings just a month ago, the distinct lack of berries by now will have initiated a mass desertion of birds. These flocks on the move will then effectively form a rolling wave trundling steadily in a south-westerly direction, so as the winter progresses, a Waxwing sighting will probably be more likely in Axminster than Aberdeen. However, with Rowan, Hawthorn, Pyracantha and Cotoneaster widespread either as native or cultivated plants right across urban, suburban and rural Britain, the only safe assumption when trying to catch up with a flock of Waxwings during an invasion year is that they could be found virtually anywhere.

Typically far more restricted to a number of long-established overwintering sites across the British Isles, a hardy bird like the Bewick's Swan will be positively revelling in the average temperature of around 4°C that typifies the month of December across Britain. Research work on the swans at WWT Slimbridge, Gloucestershire has shown that many of them are very site-faithful, and being relatively long-lived birds, many individuals are recorded returning to the same location year after year. By identifying the individual Bewick's Swans from their unique bill markings the researchers at Slimbridge have been able to calculate that 87% of the adults survive from one winter to the next, and even for the younger, more inexperienced birds, the figure is around 66%. Of all the adults identified at the reserve each winter around two-thirds are paired, and with monogamy being the norm amongst these birds, the vast majority of pairings will only be broken by the death or disappearance of a partner. In fact 'divorce' is so rare amongst Bewick's Swans that at Slimbridge mate-switching has only been recorded once while each partner was still alive. In this one exceptional case the birds 'Piotr' and 'Patch' were very much an item during the winters of 2001/02 and 2002/03, however in 2003/04 and 2004/05 'Patch' was suddenly found to have paired with 'Peploe', leaving poor old 'Piotr' to return alone. So deep-rooted is this monogamy that if one of the swans loses its mate, the researchers have calculated it could then take the surviving bird around five years to recover from becoming widowed, before breeding successfully with a new partner.

In stark contrast, the Cuckoo is highly polygynous on its breeding grounds, and so it is unlikely sexual congress will be at the forefront of their minds as they continue to see out the rest of the year in either the Congolese forests or northern Angola. In both December 2013 and December 2014, when Chris the Cuckoo headed to Angola, he initially arrived in the northern province of Uíge. This region was historically covered with forest, wooded grassland and swamps, but has since largely been converted into a rich pastoral terrain reserved primarily for farming. After initially pitching down, on both occasions Chris then quickly moved further south, to near the coastal (and capital) city of Luanda. The coastal habitat surrounding Luanda is more arid than further inland, with large areas clothed in thick scrub. Quite why Chris and a select few of the other Cuckoos would leave the sanctuary of the Congo for the drier and more broken habitat of northern Angola is anyone's guess; however, in 2013 the rainfall in the Congo was considered higher than normal. This wetter than average winter in Africa's greatest forest may simply have made feeding conditions more unfavourable for the Cuckoos, which in turn forced a few adventurous individuals to seek better foraging opportunities further south.

It was not until the data from the geolocator attached to Nightingale OAD were analysed that researchers at the BTO finally had the first definitive evidence that British-breeding birds spent at least part of their winter in the coastal scrub of West Africa. Up to that point the only record of any British Nightingale recovered in Africa was that of a bird originally ringed near Shrewton in Wiltshire in June 1974, which was subsequently found freshly dead in Morocco in April of the following year. Since the pioneering work with OAD, two British-breeding birds have been netted by British ringer Jez Blackburn and his team at Kartong Bird Observatory in The Gambia in the middle of December in 2011 and 2012. Situated on the coast and just south of the capital Banjul, the ringing location is centred around the Kartong Wetlands, a diverse site which comprises reed and rush beds, sand dunes, mudflats, mangroves, rice fields, savanna scrub and a remnant of what would have been the original forest cover. Both Nightingales were caught in exactly the same patch of dense thorny scrub, with the 2011 bird originally ringed as a juvenile near Ipswich, Suffolk in June of the same year, while the other Nightingale had previously been caught at Foulness, Essex in July 2011, only then to find itself being extracted from a second mist net, some 17 months later and 4,568km away. This remarkable connection between the counties of Suffolk and Essex, and a seemingly small and inconsequential site on the coast of west Africa, is all the more poignant given the fact that this apparently crucial scrub habitat continues to be cut down for firewood by the locals – one more reason why, with just 6,700 singing males recorded in the last survey, the Nightingale is continuing its alarming decline as a breeding bird in Britain.

Occupying around 760,000 breeding territories in Britain, the Swallow is still thankfully one of our commonest summer migrants. Those British Swallows which successfully negotiated the epic journey down to their South African wintering grounds should by now be finding the weather becoming ever more agreeable with each passing week. With conditions becoming steadily warmer and drier, resulting in an exponential increase of flying invertebrates, the Swallows will need to be wary to ensure they themselves don't also become targeted. The large winter roosts must be the equivalent of large honey pots to the likes of Eurasian Hobbies, Peregrines and Lanner Falcons as they attempt to snatch a meal during the brief time-slots when the Swallows are either descending to the roost at dusk or leaving at dawn. Having reached the sanctuary of the roost, the Swallows will still not be able to fully switch off, as Marsh Mongooses are capable of snaffling the occasional unwary Swallow. Generally though, these low levels of predation should not even scratch the surface of the large populations safely managing to roost at many sites.

Whether Puffins are subject to predation during the winter months is currently unknown, but it's a fair assumption that a combination of poor weather and starvation will surely be the biggest factors preventing these tough little seabirds from seeing the winter through. At most British colonies the Puffins will first start to be seen in the offshore waters from early March onwards, but thanks to the deployment of geolocators attached to a number of Puffins breeding on the Isle of May, it now seems that the first moves in the direction of the breeding grounds are made much earlier. In pioneering work following 13 Puffins out at sea following their dispersal from breeding grounds on the Isle of May, researcher Mike Harris and his team were amazed to discover that ten of these birds quickly left the North Sea for presumably better feeding areas in the Atlantic Ocean. However, by mid-December, all eight of the birds whose devices were still functioning, had returned to the North Sea. This suggests that their wing moult, which will have resulted in a four or five week period of flightlessness, must have finished by then, leaving just the moult into breeding plumage to be carried out. The reason for moving ever closer to the breeding colonies with a full three months yet to go can only be speculated upon, but being a species that is not only site-faithful, but also frequently burrow-faithful, might suggest that 'the early Puffin will get the burrow'!

Late December

As the nation sits down to enjoy a Christmas dinner with all the trimmings, Robins up and down the land will also be using the season of goodwill as a time to change their tune. Being the only British species to sing persistently throughout the Christmas and New Year period, the Robin's melancholic winter song, which commenced after their annual moult in midsummer, will suddenly become replaced by a much more strident version. This clear signal that spring has just come under starter's orders, will see those males holding territories suddenly declaring their occupancy from much more prominent positions amongst the bare branches. Singing at potentially any time through the day, and even at night, when some urban Robins are tricked by streetlights into thinking dawn is imminent, their characteristic song is capable of considerable variation. A number of the notes even appear to be well above our audible range, making the song sound different and possibly even more impressive to those birds within earshot.

Like humans, some Robins are far more proficient songsters than others, and those males which have not yet managed to attract the attentions of a female are generally considered to sing better and for longer. Each male will of course only sing from the comfort of his own territory and so by jotting down on a map every perch from which the bird sings, it can be a straightforward process to determine the boundaries of each territory. In fact the renowned Robin expert David Lack found it impossible to drive a singing bird from its territory, and stated 'as the observer approaches the bird retreats, but on reaching the edge of its territory it does not proceed further, and if chivvied it unexpectedly flies back over the observer's head to the middle of its ground'.

While the nation's favourite bird holds firm over the festive season, so too will any established pair of Tawny Owls. The Long-eared and Tawny are comfortably the most nocturnal of all the British owls, and so with daylight lasting for barely a third of each 24-hour period by the time the winter solstice arrives, the Tawnies should have plenty of time for hunting, territory maintenance, pair bonding and taking the first tentative steps towards the breeding season. Despite an intimate knowledge of their territory, a temporary shortage of food will surely be the biggest cause of mortality in overwintering owls, potentially paving the way for a spot of opportunism to pay dividends. Nevertheless Hamish Smith from the Hawk and Owl Trust was truly surprised when checking the camera footage from a Peregrine nest situated on a church in the city of Bath in 2015, to find one enterprising urban Tawny Owl caught repeatedly visiting the church. Presumably having flown across from tree cover close by, this brazen individual was then filmed on a number of occasions relieving the resident Peregrines of some of the contents their winter cache!

In the 'dog eat dog' world of winter survival, some urban Peregrines have also developed a new trick up their sleeve for extending the number of hours they are able to spend hunting at this time of general food shortage in the natural world. The technique of hunting nocturnal migrants lit up by the city lights below was conclusively proved in December 2010, at Derby Cathedral, when a territory-holding Peregrine was filmed bringing in a Woodcock at 10.45pm. Filmed by a static camera as the Peregrine returned to the ledge, the grainy footage clearly shows the prey still alive on being brought in, before then being dispatched with a bite to the rear of its neck – definitive proof it had been plucked out of Derby's illuminated night sky. For the resident pair of Peregrines staying close to their cathedral territory throughout that winter, the hapless Woodcock would certainly have provided a change from their normal Christmas fare of Feral Pigeon!

Any Kingfisher still holding territory come the end of the year will certainly not be 'wishing for a white Christmas', as the accompanying freezing temperatures can spell real trouble for a bird which needs water to be maintained in its liquid form if it is to catch sufficient food. At those few sites which may not have frozen over due to a quirk of topography or aspect, meaning fish are still accessible, the incumbent Kingfisher may have to accept that chasing off other starving Kingfishers is simply counter-productive. With territoriality quickly breaking down at these locations, the fishing may well descend into a free for all, until the weather relents sufficiently for the trespassers to be able to return to their own ice-free territories.

Being far more mobile than the Kingfisher, any Lapwing flocks suddenly caught out by a cold snap will simply respond with their wings by abandoning anywhere frozen for more agreeable conditions elsewhere. Often relocating on a broad front, the Lapwings will usually move during the day in a southerly or south-westerly direction as they attempt to track down frost-free food elsewhere. Occasionally these flocks can be huge, such as the 40,000 Lapwings seen along the Sussex coast near Shoreham in 1978. These large, mobile flocks will often not just consist of Lapwings, but also include Golden Plovers, Skylarks, thrushes and Starlings in their midst, all of which will be keen to tap into the Lapwings' instinctive ability for finding fair-weather feeding conditions. Due to the recent run of mild winters in Britain, these large-scale movements have become far more sporadic, but in late December 2010, for example, when virtually the whole of Britain became snowbound, it's thought the vast majority of Lapwings would have simply left for France and Spain, only returning to Britain when conditions improved.

For the Nightingales down in tropical Senegal and Gambia, plunging temperatures are highly unlikely ever to be an issue around Christmas. However, with the rainy season having already abruptly ended back in October, the steady drying out of the acacia scrub under the unrelenting sun is thought to be the decisive factor which will ultimately force the birds elsewhere come December or January. The fact that British-breeding Nightingales make a late winter movement away from Senegal and Gambia had not been realised until the tracking of a number of the birds with geolocators was started by the BTO in 2009. Nightingale OAD, and a further five other tagged birds subsequently followed in 2012, were all believed to have moved to the narrow coastal strip between Guinea-Bissau and Sierra Leone after originally setting up a temporary home up in Senegal and Gambia earlier in the winter. This region is considered to be more forested than the coastal areas further north, and with the rainy season drawing to a close marginally later in this region, the ability of the vegetation to retain that extra moisture for longer should still mean plenty of invertebrate food on offer later in the year than just up the coast. Nightingale OAD in 2009, for example, was thought to have departed Senegambia in mid to late December to spend the rest of its winter in the south-west corner of Guinea.

Just to the south of the diminutive country of Guinea-Bissau, the Republic of Guinea is roughly the size of the UK, with a population of just over 10 million. Despite being still heavily forested in some regions and with rich deposits of diamonds, gold and bauxite, the vast majority of the population works in the agricultural sector, seeing little of the country's natural wealth. Next to nothing is known about precisely which habitats the Nightingales will be spending their time in this remote and understudied part of Africa, but the remnant fragments of the Western Guinean lowland forest ecoregion or the remaining patches of mangrove forests dotted along the Atlantic coast may figure prominently. Finding enough food during this period, as they prepare for the migration back north, will be crucial if this declining British species is to defy the odds, not only by surviving, but by continuing to successfully raise another brood in our southern woods in just over four months' time.

Back in Britain, the only factor pushing the Blue Tits out of their small winter range would be the arrival of snow, but this would only represent a temporary setback for these resilient little birds, as they will then quickly return to familiar surroundings the instant the thaw sets in. Any garden with a well-stocked feeding station nearby may cause many of the local Blue Tits to distort their home range to make sure this location becomes a regular fixture on their daily rounds. These Blue Tits of course won't have a monopoly on the food available, and must also contend with hungry members from other species, in addition to other individuals from within their own clan, if they are to garner enough food to make their visit worthwhile. Amongst the tits a pecking order will frequently occur, with the bigger, bulkier Great Tits dominating the other species. Coming a close second to their bigger cousins, the pugnacious character of the Blue Tits will stand them in good stead as they then, in turn, bully the smaller and more timid Coal Tits and Long-tailed Tits off the feeders.

In particularly cold weather, well-stocked gardens can become heaving with small birds, with BTO ringer Denise

Wawman managing to count an astonishing 60 different Blue Tits at the same time in her small Somerset garden during the bitterly cold December in 2010. In this scenario, and with the stakes so high, it will surely be a case of every Blue Tit for itself, with the meek and mild potentially going hungry. As two Blue Tits compete for one perch on a feeder, each bird will attempt to assert its dominance by raising its crest, fluffing out its feathers, raising its wings and opening its beak. As the birds face off, each will then have to make a snap decision as to whether to gamble by holding its position and returning the aggression with interest, or to quickly retreat in the face of such hostility. Countless numbers of these 'mini-duels' will occur around each feeder during these cold, short days, as the winner is rewarded with a beakful of food, which will then be spirited away to be devoured out of the spotlight in a secluded bush nearby.

Having already spent over four months well out at sea, it will probably be at least a further two and a half months before the adult Puffins will even contemplate touching down back on dry land. This will not stop them steadily moving closer to their respective breeding grounds, even though they will still be too far from shore to be spotted from either cliff-tops or promontories. Presumably the location they adopt for this 'holding pattern' will contain sufficient food reserves to allow the finalisation of their moult into the full breeding plumage. As the adult clans move closer to home, this will contrast with the young puffins, which in most cases will opt to spend the summer much further away from the bustling breeding colonies. Certainly for those birds hatching during the summer, and which will be currently seeing out their first winter, the moult of their main flight feathers won't happen until the following spring and early summer. With a whole variety of predators accumulating around a puffinry in summer, this is presumably the last place a flightless and inexperienced Puffin would want to be. In reality, it may not be until the Puffins approach their third birthday that they will even be tempted back to visit the breeding colony, as these adolescents carry out their first tentative breeding reconnaissance. Recently added to the red list of British birds preserved for species of 'the highest conservation concern' due to a worldwide population decline, it must be hoped that this long-lived, iconic and characterful seabird is able to reverse this worrying trend.

Having spent four consecutive Christmases anywhere between 6,400km and 6,900km from his regular summering ground, it's remarkable to think that Chris the Cuckoo could have clocked up around 125,000km in between when he was tagged near Santon Downham, Suffolk in June 2011 and his sad demise in the deserts of northern Chad in September 2015. In addition to visiting 15 different African countries, as well as most of western Europe in his colourful life, Chris has provided researchers with a wealth of information as they attempt to tackle exactly why the Cuckoo is declining so quickly as a British-breeding bird. Having been able to follow Chris's every movement for over four years, two important facts stand out. With Chris having spent three times longer in the Congo than in Suffolk, then surely he is far more African than British? And also as a species with no regard for borders or frontiers, it needs to be an international effort right across the Cuckoo's entire migration route if we are to conserve this wonderful and mysterious species, currently celebrating either a Congolese or an Angolan Christmas.

As the end of the year approaches it will be at least another six weeks before the Swallows begin to depart South Africa for their long circuitous route back to rural Britain. With even the stragglers having arrived by December, the sheer number of Swallows skimming the South African skies will make this species surely one of the country's most populous winter visitors. The Barn Swallow is the most widespread species of Swallow in the world, and with six recognised subspecies can be seen during the appropriate season on every continent except Antarctica. Having lived alongside humans for thousands of years, it's no surprise that a bird of such cultural significance in so many countries has been decorated with so many names, such as Golondrina in Spain, the bird that thaws the snow; Svala in Sweden, meaning console; or perhaps the most evocative of all, Nyankalema in Zambia, translating as the bird that never gets tired.

Dedicated to the memory of celebrated artist and ornithologist Thomas Bewick, Bewick's Swans should be reaching their maximum counts in Britain around the turn of the year. Having fed well on our arable crops, pasture lands and estuaries, the birds will also be at their heaviest weight since their arrival in late October. With most of the fat stored between the legs and tail, researchers are able to see which birds have fed particularly well, with Julia Newth of WWT Slimbridge saying, 'in a slim bird the bum will look slightly concave, whereas a well-fed bird will have a double-bulge'. This would suggest that birds with 'big healthy behinds' may fare better when the time comes to depart for their breeding grounds in late February. During the last census of Bewick's Swans in January 2005, a grand total of 7,216 Bewick's Swans were recorded at just 26 sites across Britain, representing 33% of all the 21,500 birds estimated to have been overwintering in north-west Europe during the same survey period. Falling from the peak count of 29,277 during the previous survey exactly a decade earlier, research work has still been unable to pinpoint exactly why this charismatic and hardy winter visitor from the Russian Arctic continues to decline.

During those years when Britain is lucky enough to be graced with a 'Waxwing winter', as birds coalesce at good feeding sites, the flock sizes can quickly build to impressive proportions, like the 1,400 seen at Pegwell Bay, on Kent's east coast in December 2010. This spectacle must have been all the more thrilling given how approachable Waxwings always seem to be on their wintering grounds. They are relatively easily to count, as they trill away among the bare branches or line up on TV aerials in between feeding bouts, but estimating the grand total of Waxwings visiting Britain in an invasion year is a much more difficult task given the mobile nature of the flocks. The best attempt at assessing Waxwing numbers has been made by BirdTrack, an online website managed by a variety of conservation organisations, which collates records sent in by 'citizen scientists' across Britain and Ireland. Reported from 4,569 1km squares during the winter of 2010/11 and from 4,980 1km squares during the last invasion year of 2012/13, this sporadic and gentle winter visitor will surely delight whomever it meets and charm wherever it goes.

Further Reading

Balmer, D., Gillings, S., Caffrey, B., Swann, B., Downie, I. & Fuller, R. 2013. *Bird Atlas 2007-2011*. BTO Books, Thetford.

Chandler, David & Llewellyn, Ian. 2010. *Kingfisher*. New Holland Publishers.

Clare, Horatio. 2009. *A Single Swallow*, Vintage, Random House, London.

Cramp, Stanley (Editor). 1985. *Handbook of the Birds of Europe, the Middle East and North Africa: The Birds of the Western Palearctic. Volume 4*. Oxford University Press, Oxford.

Cramp, Stanley (Editor). 1988. *Handbook of the Birds of Europe, the Middle East and North Africa: The Birds of the Western Palearctic. Volume 5*. Oxford University Press, Oxford.

Davies, N. B. 2000. *Cuckoos, Cowbirds and other Cheats*. T & A.D Poyser, London.

Davies, Nick. 2015. *Cuckoo - Cheating by Nature*. Bloomsbury Publishing, London.

Drewitt, Ed. 2014. *Urban Peregrines*. Pelagic Publishing, Exeter.

Ferguson-Lees, J., Castell, R., & Leech, D. 2011. *A Field Guide to Monitoring Nests*. The British Trust for Ornithology, Thetford.

Flegg, Jim. 1987. *The Blue Tit*. Shire Publications, Bucks.

Ginn, H. B. & Melville, D. S. 1983. *Moult in Birds*. The British Trust for Ornithology, Thetford.

Hamilton James, Charlie. 2009. *Kingfisher - Tales from the Halcyon River*. Evans Mitchell Books.

Harris, Mike & Wanless, Sarah. 2011. *The Puffin*. T & A.D Poyser, London.

Holden, Peter & Cleeves, Tim. 2014. *RSPB Handbook of British Birds, 4th Edition*. Bloomsbury Publishing Plc., London.

Lack, David. 1965. *The Life of the Robin*. H.F & G. Witherby, London.

Mead, Chris. 1984. *Robins*. Whittet Books, London.

Mead, Chris. 1987. *Owls*. Whittet Books, London.

Pike, Oliver, G. 1932. *The Nightingale - Its Story and Song*. J.W. Arrowsmith, Bristol.

Ratcliffe, Derek. 1993. *The Peregrine Falcon.* T & A.D Poyser, London.
Rees, Eileen. 2006. *Bewick's Swan.* T & A.D. Poyser, London.
Shrubb, Michael. 2007. *The Lapwing.* T & A.D Poyser, London.
Snow, Barbara & David. 1988. *Birds and Berries.* T & A.D. Poyser, London.
Tate, Peter. 1986. *The Swallow.* Shire Publications, Bucks.
Toms, Mike. 2003. *The BTO/CJ Garden BirdWatch Book.* The British Trust for Ornithology, Thetford.
Toms, Mike, 2014. *Owls.* William Collins, London.
Turner, Angela. 2006. *The Swallow.* T & A.D Poyser, London.
Taylor, Kenny. 1993. *Puffins.* Whittet Books, London.
Wernham, C., Toms, M., Marchant, J., Clark, J., Siriwardena, G., & Baillie, S. 2002. *The Migration Atlas: Movements of the Birds of Britain and Ireland.* T & A.D. Poyser, London.
Wyllie, Ian. 1987. *The Cuckoo.* Shire Publications, Bucks.

Acknowledgements

First and foremost thanks to my wonderful partner Christina, and not just for her constant support, but also for taking on the majority of domestic and child-minding duties while I either disappeared into my small, shambolic office, or was away filming. Huge appreciation must also go to the Dilger and Holvey families, particularly Renee, Graham and Laura who frequently go above and beyond in the 'familial duties' that might normally be expected. My agents Hilary Knight, Phyllida Knight and Jane Turnbull always have my back, and Jane particularly was instrumental in ensuring not only that this book was commissioned in the first place, but that it was also finished!

If birds fascinate you, and you are not yet a member of the British Trust for Ornithology, then I implore you to join. This charity is a shining beacon for both 'Citizen Science' and putting all their hard-won data into the public forum. Their online resources are quite simply staggering and everyone I have come across at the charity has been only too happy to help me source information and respond to endless queries with unrelenting enthusiasm. Thanks particularly must go to Paul Stancliffe, Carole Showell, Chris Hewson, Jez Blackburn, Andy Clements, Carl Barimore and Mike Toms. I'm also a big supporter of the Wildfowl and Wetlands Trust, and James Lees, Dave Paynter and Larry Griffin have been incredibly generous in furnishing me with invaluable information and first-hand accounts.

Outside the conservation sector I have also been fortunate to have dipped into a substantial number of wells of expertise. Denise Wawman has helped provide colourful insight into the private lives of her Robins and Blue Tits, Terry Holvey has shared his obsession with Blue Tits, Charlie Hamilton James' knowledge on Kingfishers is world-class and what my good pals Ed Drewitt and Dave Culley don't know about Peregrines and Tawny Owls respectively is frankly hardly worth knowing. I'm also indebted to Hamish Smith for filling in some of the blanks about Peregrines, Juan Brown for helping with Puffins and Raymond Duncan for passing on so much interesting data on Waxwings.

At Bloomsbury I'm grateful to Jim Martin for not only steering me in the right direction but also gently prodding me at the right time, Jane Lawes for overseeing the production of the book with seamless efficiency and Sara Hulse for her sharp, incisive and grammatically spot-on edit. I was also delighted when Darren Woodhead agreed to carry out the artwork. His deft brush and pencil strokes bring the birds to life, making them seemingly fly straight off the page, and the book is immeasurably better for his contribution.

Finally, I would like to place on record my immense gratitude to two very important people who have had a strong guiding hand in shaping my career over the last decade. Stephen Moss and Nigel Redman are not just brilliant birders, but they are also great friends - and I thank you gentlemen for your comradeship, advice and generosity of spirit.

Index

abmigration 59–60
Anemone, Wood 99
apples 24, 45
April 100–101
 Cuckoo 106–108
 Kingfisher 102–103, 117–18, 121–22
 Lapwing 103–104, 117, 118, 119–20
 Nightingale 108, 110–11, 126–27
 Owl, Tawny 101–102, 115–16, 122–23
 Peregrine 103, 116–17, 120–21
 Puffin 105–106
 Robin 104–105, 116, 123–24
 Swallow 100–101
 Swan, Bewick's 109, 114–15, 128
 Tit, Blue 108–109, 118–19, 124
 Waxwing 109–10, 115, 127–28
August 223
 Cuckoo 225–26, 227, 234–35, 246–47
 Kingfisher 228–29, 239, 243
 Lapwing 231, 239–40
 Nightingale 224–25, 227, 233–34, 247–48
 Owl, Tawny 232–33, 238–39, 243–44
 Peregrine 230–31, 237–38, 244–45
 Puffin 226–27, 233, 245–46
 Robin 230, 240–41, 243
 Swallow 227–28, 235–36, 248
 Swan, Bewick's 229–30, 236–37, 242
 Tit, Blue 231–32, 240, 242–43
 Waxwing 230, 237, 241–42
Badger 118, 129
Bearberry 197
Beech 99
Bempton Cliffs, Yorkshire 144–45
Bilberry 133, 196
Birch 133
Bittern 99
Blackbird 26, 45, 70, 123, 179, 266
Blackcap 99, 231
Blenny 292
Bluebell 99
Boar, Wild 340
Bonxie 206
British Trust for Ornithology (BTO) 10, 29, 30, 31, 39, 40, 55, 59–60, 80, 85, 106, 107, 127, 131, 171, 191, 213, 220–21, 224, 225, 232, 250, 253, 341, 354, 355
Bullhead 185, 292
Buzzard 149
calls
 Cuckoo 130–31
 Owl, Tawny 26, 284, 293, 300–301
Carp 185
Celandine, Lesser 71
chicks
 Cuckoo 161–62, 172–73, 177–78, 191–92, 203–204, 226, 246–47
 Kingfisher 121–22, 137–38, 151–52, 153–54, 170–71, 172, 185, 203, 218–19, 228–29
 Lapwing 119–20, 121, 135–36, 149, 154, 167–68
 Nightingale 140, 162–63, 168–69, 183
 Owl, Tawny 101–102, 115–16, 122–23, 136, 149–50, 155, 172, 178–79, 190, 199, 208, 216, 232–33
 Peregrine 116–17, 120–21, 136–37, 150–51, 155–56, 167, 179, 190–91, 199–200, 207–208, 221–22, 230–31, 237–38
 Puffin 176–77, 192–93, 206, 214–16
 Robin 116, 123–24, 134–35, 148–49, 154–55, 169, 170, 172, 189, 201–202
 Swallow 173–74, 180–82, 187–89, 202–203, 208–209, 219, 235–36
 Swan, Bewick's 196, 205–6
 Tit, Blue 116, 152–53, 156–58, 166–67, 200, 240, 242–43
 Waxwing 192–93, 196–97, 204–205
Chiffchaff 72, 100, 231
Cloudberry 133, 229
Cod 177
Corncrake 117, 291
Cotoneaster 19, 45, 54, 346
Cowberry 133, 197, 229
Crow, Carrion 63, 118, 149
Crowberry 196, 229
Cuckoo 10, 11, 15
 April 106–108, 113, 127
 August 225–26, 227, 234–35, 246–47
 December 335–36, 347–48, 357
 February 55, 61–62, 68
 January 30–31, 39–40, 45–46
 July 203–204, 213–14, 220–21
 June 172–73, 177–78, 191–92
 March 78–79, 86, 91–92
 May 130–32, 141–44, 161–62
 November 312, 317–18
 October 276–77, 278, 286–87, 289, 297–98
 September 253, 264, 272
Cuckoo Chris 30–31, 86, 92, 106–109, 113, 127, 191–92, 213, 221, 234, 246, 253, 271, 276–77, 287, 335, 347–48, 357
Culley, Dave 57, 72, 94, 123, 179, 208
Daffodil 71
Darwin, Charles 266
Davies, Nick 143, 162
December 333
 Cuckoo 335–36, 346–48, 357
 Kingfisher 338, 343–44, 353
 Lapwing 338–39, 345–46, 353–54
 Nightingale 336, 348–49, 354–55
 Owl, Tawny 334, 344–45, 351–52
 Peregrine 340–41, 343, 351–53
 Puffin 350, 356–57
 Robin 339–40, 342–43, 350–51
 Swallow 334–35, 349–50, 357–58
 Swan, Bewick's 336–37, 346–47, 358–59

Tit, Blue 339, 341–42, 355–56
Waxwing 337–38, 346, 359
Deer, Red 275
diet
Cuckoo 45–46, 131
Kingfisher 122, 137–38, 185, 292
Lapwing 43–44, 149, 301–302, 343–44
Owl, Tawny 35–36, 65–66, 94, 123, 256, 322–23, 344–45
Peregrine 37–38, 117, 321
Puffin 46–47, 176–77, 197–99, 320–21
Robin 36, 149, 321–22
Swallow 31–32, 47, 181, 227–28
Swan, Bewick's 23, 44, 53, 109, 319
Waxwing 23–24, 33, 45, 53–54, 115, 133, 196–97, 319–20
Dipper 266
Dixon, Nick 321
Dormouse, Edible 322
Dove, Turtle 117, 290
Drewitt, Ed 37, 42, 81, 117, 254, 266, 291, 321
Dunlin 316
Dunnock 131, 141
eggs
Cuckoo 143–44
Kingfisher 102, 104, 117–18, 189–90
Lapwing 103–104, 118
Nightingale 140, 144
Owl, Tawny 57–58, 72–73, 81–82, 94, 95, 104
Peregrine 81, 94–95, 103, 104
Puffin 139, 144–45, 164, 174
Robin 104–105
Swallow 146–47, 163–64
Swan, Bewick's 174–76, 183–84
Tit, Blue 118–19, 124, 138–39
Waxwing 184–85
Falcon, Lanner 349
February 49–50
Cuckoo 55, 61–62, 68
Kingfisher 49, 52, 58–59, 62–63
Lapwing 49, 52–53, 59–60, 63–64
Nightingale 61–62, 70
Owl, Tawny 49, 51–52, 57–58, 65–66
Peregrine 49–50, 55–56, 64–65
Puffin 54, 60–61, 68–69
Robin 49, 50, 57, 66
Swallow 54–55, 60, 67–68
Tit, Blue 49, 50–51, 56–57, 66–67
Waxwing 53–54, 61, 70
Fieldfare 45, 70, 321
flocking
Lapwing 28–29, 43, 52–53, 183, 186, 306–308, 353–54
Puffin 92–93
Tit, Blue 26–27, 50, 231–32, 265, 289–90, 299–300
Waxwing 33–34, 285, 296–97, 359
Fox 16, 26, 63, 81–82, 118, 129, 345
Arctic 184, 205
Garlic, Wild 99
geolocators 29–30, 32, 39–40, 61–62, 85–86, 108, 224, 232, 245, 247, 253, 263, 272, 298, 311, 313
Goby 291
Goldcrest 27, 231, 232, 265, 321
Gorse 129
Grayling 185
Greenfinch 342
Guelder Rose 45
Guillemot 105, 114, 139, 198
Gull, Black-headed 303
Common 303
Great Black-backed 93, 206
Herring 65, 184, 207
Lesser Black-backed 207
Haddock 177
Harris, Mike 32, 197, 272, 298, 330, 350
Hawk and Owl Trust 352
Hawthorn 24, 45, 99, 346
Herring 176
Hobby 209, 349
Hosking, Eric 82
Hudson, William Henry 266
James, Charlie Hamilton 102, 239
January 21–22
Cuckoo 30–31, 39–40, 45–46
Kingfisher 27–28, 34, 43
Lapwing 28–29, 34–35, 43–44
Nightingale 29–30, 46
Owl, Tawny 25–26, 35–36, 40–41
Peregrine 27, 37–38, 42–43
Puffin 21–22, 32, 38, 46–47
Robin 25, 26, 27, 36, 41–42
Swallow 31–32, 38–39, 47
Swan, Bewick's 22–23, 32–33, 44
Tit, Blue 26, 36–37, 42
Waxwing 23–25, 33–34, 44–45
Jay 345
July 195
Cuckoo 203–204, 213–14, 220–21
Kingfisher 203, 209–11, 218–19
Lapwing 201, 211, 212, 222
Nightingale 200–201, 211, 212–13, 220
Owl, Tawny 199–200, 208, 216
Peregrine 207–208, 221–22
Puffin 197–99, 206–207, 214–16
Robin 201–202, 211, 219–20
Swallow 202–203, 208–209, 219
Swan, Bewick's 196, 205–206, 216–17
Tit, Blue 200, 211–12, 219–20
Waxwing 196–97, 204–205, 217–18
June 165–66
Cuckoo 172–73, 177–78, 191–92
Kingfisher 170–72, 185, 189–90
Lapwing 167–68, 183, 186
Nightingale 168–69, 183, 186–87
Owl, Tawny 172, 178–79, 190
Peregrine 167, 179, 190–91
Puffin 174, 176–77, 192–93
Robin 169–70, 179–80, 189
Swallow 173–74, 180–82, 187–89
Swan, Bewick's 174–76, 183–84, 194
Tit, Blue 166–67, 182–83, 187
Waxwing 174–75, 184–85, 193–94
Kingfisher 16–17, 104
April 102–103, 117–18, 121–22
August 228–29, 239, 243
December 338, 343–44, 353
February 49, 52, 58–59, 62–63
January 27–28, 34, 43
July 203, 209–11, 218–19
June 170–72, 185, 189–90
March 73–74, 82, 96
May 137–38, 151–52, 153–54
November 315, 323–24, 324–25

October 282, 291–92, 302
September 255–56, 260, 265–66
Knot 223, 316
Lack, David 41, 83, 256, 326, 351
Lanternfish, Glacier 321
Lapwing 14, 112, 121
April 103–104, 117, 118, 119–20
August 231, 239–40
December 338–39, 345–46, 353–54
February 49, 52–53, 59–60, 63–64
January 28–29, 34–35, 43–44
July 201, 211, 212, 222
June 167–68, 183, 186
March 74–75, 84, 95–96
May 135–36, 149, 150, 154
November 306–308, 324, 326
October 281–82, 289, 302–303
September 250–51, 261, 269
Lees, James 69
Lichen, Witch-hair 133
lifespans
Kingfisher 171
Peregrine 55, 266–67
Puffin 61
Robin 179–80
Swan, Bewick's 23, 327
Waxwing 325–26
Loach 185
Mallard 321
March 71–72
Cuckoo 78–79, 86, 91–92
Kingfisher 73–74, 82, 96
Lapwing 74–75, 84, 95–96
Nightingale 78, 85–86, 91
Owl, Tawny 72–73, 81–82, 94, 95
Peregrine 73, 80–81, 94–95
Puffin 77–78, 85, 92–93
Robin 75–76, 82–83, 97
Swallow 79–80, 87–88, 89–91
Swan, Bewick's 76–77, 89, 93–94
Tit, Blue 75–76, 83–84, 97
Waxwing 77, 88–89, 93–94
Martin, Sand 72, 100, 112
mating
Kingfisher 96
Lapwing 75, 95–96
Peregrine 73
Puffin 124–25
Swallow 125–26, 145–46
Swan, Bewick's 159–60
Waxwing 160–61
May 129–30
Cuckoo 130–32, 141–44, 161–62
Kingfisher 137–38, 151–52, 153–54
Lapwing 135–36, 149, 150, 154
Nightingale 139–40, 144, 162–63
Owl, Tawny 136, 149–50, 155
Peregrine 136–37, 150–51, 155–56
Puffin 139, 144–45, 164
Robin 134–35, 148–49, 154–55
Swallow 140–41, 145–47, 163–64
Swan, Bewick's 133–34, 147, 158–60
Tit, Blue 138–39, 152–53, 156–58, 162
Waxwing 132–33, 147–48, 160–61
Merlin 218
migration 9–10
Cuckoo 30–31, 68, 78–79, 86, 91–92, 106–108, 113, 127, 213–14, 220–21, 225, 234–35, 276–77
Lapwing 28–29, 35, 43, 59–60, 186, 222, 231, 250–51
Nightingale 29–30, 70, 78, 85–86, 87, 108, 224, 233–34, 253–54, 263–64, 271–72, 278, 297
Peregrine 260–61, 290–91
Puffin 68–69, 77–78, 226–27, 264–65, 272, 298–99, 312–13
Swallow 54–55, 60, 67–68, 87–88, 89–91, 100–11, 248, 251–52, 262–63, 272–74, 277, 293–94
Swan, Bewick's 69–70, 76–77, 89, 93–94, 114–15, 128, 133–34, 147, 158–59, 252, 269–70, 286
Tit, Blue 36–37, 79–80, 279–80
Waxwing 77, 88–89, 93–94, 109–10, 127–28, 252–53
Mink, American 102–103
Minnow 185, 291
Mistletoe 45
Moorhen 37
Moreau, Reginald 276
Moth, Winter 157
moulting
Cuckoo 270–71, 317–18
Kingfisher 43, 210–11, 218–19, 243, 282, 324–25
Lapwing 53, 168, 201, 211, 281–82
Nightingale 187, 200–201, 211, 212–13
Owl, Tawny 172, 243–44, 334
Peregrine 282–83, 300
Puffin 38, 54, 233, 283, 290
Robin 211, 219–20, 230, 283–84
Swallow 38–39, 67, 278, 318
Swan, Bewick's 216–17, 236–37, 279, 318–19
Tit, Blue 182–83, 187, 200, 201, 211–12, 219–29
Waxwing 230
Mouse, Wood 35, 65, 139, 322, 344
nest boxes 40–41, 57, 80–81, 83, 95, 119
nesting
Cuckoo 131–32, 141–44
Kingfisher 73–74, 171–72
Lapwing 74–75, 96
Nightingale 139–40
Peregrine 80–81
Puffin 105–106, 113–14
Robin 82–83
Swallow 112, 140–41
Tit, Blue 83–84, 108–109
Waxwing 174–75
Nightingale 11, 12, 18, 50
April 108, 110–11, 126–27
August 224–25, 227, 233–34, 247–48
December 336, 348–49, 354–55
February 61–62, 70
January 29–30, 46
July 200–201, 211, 212–13, 220
June 168–69, 183, 186–87
March 78, 85–86, 91
May 139–40, 144, 162–63
Nightingale 98, 86
Nightingale OAD 29–30, 39–40, 85, 224, 247–48, 263–64, 271, 311, 316, 348–49, 354
November 310–12, 316–17, 329
October 278, 288–89, 297
September 253–54, 263–64, 271–72
Nightjar 117
November 305–306
Cuckoo 312, 317–18
Kingfisher 315, 323–24, 324–25
Lapwing 306–308, 324, 326
Nightingale 310–12, 316–17, 329

Owl, Tawny 313–14, 322–23, 329
Peregrine 315–16, 321, 329–30
Puffin 312–13, 320–21, 330
Robin 314–15, 321–22, 328–29
Swallow 310, 318–19, 330–32
Swan, Bewick's 308–309, 319, 327–28
Tit, Blue 315, 323, 326–27
Waxwing 309–310, 319–320, 325–26
oaks 99
October 275
Cuckoo 276–77, 278, 286–87, 289, 297–98
Kingfisher 282, 291–92, 302
Lapwing 281–82, 289, 302–303
Nightingale 278, 288–89, 297
Owl, Tawny 284, 292–93, 300–301
Peregrine 282–83, 290–91, 300
Puffin 283, 290, 298–99
Robin 283–84, 292, 301–302
Swallow 277–78, 287–88, 289, 293–94
Swan, Bewick's 278–79, 286, 294–96
Tit, Blue 279–80, 289–90, 299–300
Waxwing 279, 285, 296–97
Osprey 317
Owl, Barn 268
Long-eared 351–52
Short-eared 268
Owl, Tawny 10, 11, 16, 104
April 101–102, 115–16, 122–23
August 232–33, 238–39, 243–44
December 334, 344–45, 351–52
February 49, 51–52, 57–58, 65–66
hunting technique 268–69
January 25–26, 35–36, 40–41
July 199–200, 208, 216
June 172, 178–79, 190
March 72–73, 81–82, 94, 95
May 136, 149–50, 155
November 313–14, 322–23, 332
October 284, 292–93, 300–301
September 254–55, 259, 268–69
Packham, Chris 30
pairing
Kingfisher 49, 58–59, 62–63, 82, 260
Lapwing 49, 63–64
Owl, Tawny 49, 259
Peregrine 42–43, 49, 50, 56, 64–65
Puffin 85, 114
Robin 41–42, 49, 66
Swan, Bewick's 346–47
Tit, Blue 42, 49
Waxwing 147–48
Peewit 14
Perch 185
Peregrine 13–14, 93, 104, 206–207, 349
April 103, 116–17, 120–21
August 230–31, 237–38, 244–45
December 340–41, 343, 352–53
February 49–50, 55–56, 64–65
January 27, 37–38, 42–43
July 207–208, 221–22
June 167, 179, 190–91
March 73, 80–81, 94–95
May 136–37, 150–51, 155–56
November 315–16, 321, 329–30
October 282–83, 290–91, 300
September 254, 260–61, 266–68
stoop dive 267–68
phenology 10
Pigeon, Feral 321, 344, 345, 353
Pike 185
Pine 133
Pipit, Meadow 15, 131, 141, 143, 162, 173, 177–78
Plover, Golden 303
Green 14
Primrose 71
Privet 45
Puffin 11, 14–15
April 105–106, 113–14, 124–25
August 226–27, 233, 245–46
December 350, 356–57
February 54, 60–61, 68–69
January 21–22, 32, 38, 46–47
July 197–99, 206–207, 214–16
June 174, 176–77, 192–93
March 77–78, 85, 92–93
May 139, 144–45, 164
November 312–13, 320–21, 330
October 283, 290, 298–99
September 254, 264–65, 272
Pyracantha 19, 346
Rabbit 113
Rail, Water 291, 341
Razorbill 105, 114, 139, 198
Redwing 45, 70, 321
ringing
Cuckoo 62–63
Nightingale 29, 62–63, 310–311
Owl, Tawny 232
Puffin 245–46
Robin 328–29
Swallow 31–32, 310
Tit, Blue 341–42, 355–56
Waxwing 133, 309–310
Robin 10, 11, 17–18, 50, 104, 116, 131, 141
April 104–105, 116, 123–24
August 230, 240–41, 243
December 339–40, 342–43, 350–51
February 49, 50, 57, 66
January 25, 26, 27, 36, 41–42
July 201–202, 211, 219–20
June 169–70, 179–80, 189
March 75–76, 82–83, 97
May 134–35, 148–49, 154–55
November 314–15, 321–22, 328–29
October 283–84, 292, 301–302
September 256–57, 257–59, 266
Rockling 199
rosehips 24, 45
Rowan 19, 24, 45, 241, 285, 296, 346
Royal Society 161
Royal Society for the Protection of Birds (RSPB) 10, 145
Bempton Cliffs, Yorkshire 144–45
Big Garden Birdwatch 26
Sandeel, Lesser 47, 176–77, 192, 199, 226
Sanderling 223
satellite tracking 15, 30, 39, 55, 61–62, 68, 76–79, 86, 89, 91–92, 93–94, 106, 113, 232, 286–87
Scott, Sir Peter 294–95
Seal, Grey 275

September 249
  Cuckoo 253, 264, 272
  Kingfisher 255–56, 260, 265–66
  Lapwing 250–51, 261, 269
  Nightingale 253–54, 263–64, 271–72
  Owl, Tawny 254–55, 259, 268–69
  Peregrine 254, 260–61, 266–68
  Puffin 254, 264–65, 272
  Robin 256–57, 257–59, 266
  Swallow 251–52, 262–63, 272–74
  Swan, Bewick's 252, 261–62, 269–70, 272
  Tit, Blue 257, 260, 265
  Waxwing 252–53, 262, 270–71
Shearwater, Manx 113
Skua, Arctic 184, 207
  Great 93, 206
Smelt, Sand 292
Snipe, Common 321
song
  Nightingale 110–11, 126–27, 220
  Robin 25, 51, 57, 66, 220, 240–41, 257–59, 266, 350–51
  Swallow 111–12, 125–26
  Tit, Blue 51, 56
Sparrow, House 345
  Song 66
Sparrowhawk 19, 27, 155, 166–67, 170, 200, 218, 232, 248
Sprat 176, 192
Springwatch 245
Spruce 133
Squirrel, Grey 139, 322
Starling 266, 342
Stickleback 185, 210, 292
Stoat 118, 135, 149, 155, 166
Swallow 10, 17
  April 100–101
  August 227–28, 235–36, 248
  December 334–35, 349–50, 357–58
  February 54–55, 60, 67–68
  January 31–32, 38–39, 47
  July 202–203, 208–209, 219
  June 173–74, 180–82, 187–89
  March 79–80, 87–88, 89–91
  May 140–41, 145–47, 163–64
  November 310, 318–19, 330–32
  October 277–78, 287–88, 289, 293–94
  September 251–52, 262–63, 272–74
Swallow, Greater Striped 293
  Pearl-breasted 293
  White-throated 293
Swan, Bewick's 13
  April 109, 114–15, 128
  August 229–30, 236–37, 242
  bill patterns 294–95
  December 336–37, 346–47, 358–59
  February 53, 58, 61, 69–70
  January 22–23, 32–33, 44
  July 196, 205–206, 216–17
  June 174–76, 183–84, 194
  March 76–77, 89, 93–94
  May 133–34, 147, 158–60
  November 308–309, 319, 327–28
  October 278–79, 286, 294–96
  September 252, 261–62, 269–70, 272
Swan, Mute 175, 205
Swift 312
Teal 37, 316
Tern, Sandwich 317
territories
  Kingfisher 27–28, 34, 52, 239, 255–56, 265–66, 302, 315
  Lapwing 59, 63, 64, 84
  Nightingale 126–27
  Owl, Tawny 35, 40–41, 51–52, 238–39, 313–14
  Peregrine 27, 244–45, 315–16
  Robin 41, 57, 75–76, 256–57, 266, 301–302, 314–15
  Swallow 113
  Tit, Blue 51, 75–76, 257
Thrush, Mistle 33, 45, 70
  Song 45, 70
Tit, Blue 10, 11, 12, 19, 116
  April 108–109, 118–19, 124
  August 231–32, 240, 242–43
  December 339, 341–42, 355–56
  February 49, 50–51, 56–57, 66–67
  January 26, 36–37, 42
  July 200, 211–12, 219–20
  June 166–67, 182–83, 187
  March 75–76, 83–84, 97
  May 138–39, 152–53, 156–58, 162
  November 315, 323, 326–27
  October 279–80, 289–90, 299–300
  September 257, 260, 265
Tit, Coal 231, 232, 355
  Great 27, 231, 265, 355
  Long-tailed 232, 355
Tortrix, Green Oak 157
Treecreeper 27, 36, 231, 265
Turnstone 223
Vole, Bank 35, 65, 322, 344
Wagtail, Pied 132, 141, 339
Wanless, Sarah 197, 198
Warbler, Dartford 36
  Reed 15, 131, 141–44, 162, 172–73, 177–78, 317
  Willow 99
Waxwing 18–19
  April 109–10, 115, 127–28
  August 230, 237, 241–42
  December 337–38, 346, 359
  February 53–54, 61, 70
  January 23–25, 33–34, 44–45
  July 196–97, 204–205, 217–18
  June 174–75, 184–85, 193–94
  March 77, 88–89, 93–94
  May 132–33, 147–48, 160–61
  November 309–310, 319–320, 325–26
  October 279, 285, 296–97
  September 252–53, 262, 270–71
Weasel 135, 166
Wheatear 72, 100, 117, 291
White, Gilbert 227
Whitebeam 45
Whiting 177
Wildfowl and Wetlands Trust (WWT) 58, 89, 93, 115
  Slimbridge 22–23, 33, 44, 69, 76, 294–95, 308, 336, 358
  Welney 337
wintering grounds
  Cuckoo 30–31, 39, 55, 286–87, 297–98, 335–36
  Nightingale 39–40, 55, 316–17, 336, 354–55
  Swallow 31–32, 55, 287–88, 330–32
  Swan, Bewick's 336–37
Winterwatch 35–36
Wolverine 184
Woodcock 37, 291, 321, 341, 352–53
Woodlark 99
Woodpecker, Great Spotted 139
Wren 27, 231, 265